Adamu Usman Mohammed

Human Exposure to Arsenic and Other Potentially Toxic Metals in Some Waters of Biu Volcanic Province, North-Eastern Nigeria

The effect of leaching from rocks into surrounding waters

Anchor Compact

Mohammed, Adamu Usman: Human Exposure to Arsenic and Other Potentially Toxic
Metals in Some Waters of Biu Volcanic Province, North-Eastern Nigeria: The effect of
leaching from rocks into surrounding waters. Hamburg, Anchor Academic Publishing
2013
Original title of the thesis: TRACE ELEMENTS HYDROGEOCHEMISTRY IN SURFACE AND
GROUND WATERS OF SOME PART OF BIU VOLCANIC PROVINCE, NORTH-EASTERN
NIGERIA: HUMAN HEALTH IMPACT

Buch-ISBN: 978-3-95489-176-4
PDF-eBook-ISBN: 978-3-95489-676-9
Druck/Herstellung: Anchor Academic Publishing, Hamburg, 2013
Additionally: Diplomica Verlag GmbH, Hamburg, Deutschland, Masterarbeit,

Bibliografische Information der Deutschen Nationalbibliothek:
Die Deutsche Nationalbibliothek verzeichnet diese Publikation in der Deutschen
Nationalbibliografie; detaillierte bibliografische Daten sind im Internet über
http://dnb.d-nb.de abrufbar

Bibliographical Information of the German National Library:
The German National Library lists this publication in the German National Bibliography.
Detailed bibliographic data can be found at: http://dnb.d-nb.de

All rights reserved. This publication may not be reproduced, stored in a retrieval system
or transmitted, in any form or by any means, electronic, mechanical, photocopying,
recording or otherwise, without the prior permission of the publishers.

Das Werk einschließlich aller seiner Teile ist urheberrechtlich geschützt. Jede Verwertung
außerhalb der Grenzen des Urheberrechtsgesetzes ist ohne Zustimmung des Verlages
unzulässig und strafbar. Dies gilt insbesondere für Vervielfältigungen, Übersetzungen,
Mikroverfilmungen und die Einspeicherung und Bearbeitung in elektronischen Systemen.

Die Wiedergabe von Gebrauchsnamen, Handelsnamen, Warenbezeichnungen usw. in
diesem Werk berechtigt auch ohne besondere Kennzeichnung nicht zu der Annahme,
dass solche Namen im Sinne der Warenzeichen- und Markenschutz-Gesetzgebung als frei
zu betrachten wären und daher von jedermann benutzt werden dürften.

Die Informationen in diesem Werk wurden mit Sorgfalt erarbeitet. Dennoch können
Fehler nicht vollständig ausgeschlossen werden und die Diplomica Verlag GmbH, die
Autoren oder Übersetzer übernehmen keine juristische Verantwortung oder irgendeine
Haftung für evtl. verbliebene fehlerhafte Angaben und deren Folgen.

Alle Rechte vorbehalten

© Anchor Academic Publishing, ein Imprint der Diplomica® Verlag GmbH
http://www.diplom.de, Hamburg 2013
Printed in Germany

Contents

CHAPTER ONE: GENERAL INTRODUCTION

1.1 INTRODUCTION 9
1.2 LOCATION, EXTENT AND ACCESSIBILTY 11
1.3 RELIEF AND DRAINAGE 11
1.4 CLIMATE AND VEGETATION 13
1.5 SETTLEMENT AND LAND USE 14

CHAPTER TWO: LITERATURE REVIEW

2.1 INTRODUCTION 16
2.2 EFFECTS OF TRACE ELEMENTS IN VOLCANIC AREAS 16

CHAPTER THREE: DETAILED GEOLOGY AND HYDROGEOLOGY OF THE STUDY AREA

3.1 DETAILED GEOLOGY 20
   3.1.1 INTRODUCTION 20
   3.1.2 PORPHYRITIC AND AMYGDALOIDAL VARIETY 23
   3.1.3 LARGE PHENOCRYSTS OF ZONED LABRADORITE 24
   3.1.4 AGGLOMERATES, TUFFS AND BROWNISH RED BASALTIC SCORIA 24
3.2 HYDROGEOLOGY OF THE STUDY AREA 27
   3.2.1 INTRODUCTION 27
   3.2.2 SURFACE WATER 28
      3.2.1.1 LAKE TILA 28
      3.2.1.2 PERENNIAL STREAM WATER 29
      3.2.1.3 GROUNDWATER IN THE BASALTS 30
   3.2.3 DATA COLLECTION 32
   3.2.4 DATA PROCESSING 32
   3.2.5 INTERPRETATION OF GROUNDWATER MAP 32
   3.2.6 STRUCTURES 34
   3.2.7 JOINTS 35

CHAPTER FOUR: HYDROGEOCHEMISTRY

- 4.1 INTRODUCTION 37
- 4.2 METHODOLOGY 37
  - 4.2.1 SAMPLE COLLECTION AND PRESERVATION 37
  - 4.2.2 WATER SAMPLE PREPARATIONS 41
  - 4.2.3 SOIL SAMPLE PREPARATIONS 41
- 4.3 ANALYTICAL TECHNIQUE 41

CHAPTER FIVE: PRESENTATION OF RESULTS/DISCUSSIONS

- 5.1 PRESENTATION OF RESULTS 45
  - 5.1.1 Water Sample 45
  - 5.1.2 MAJOR ELEMENTS 82
    - 5.1.2.1 Calcium (Ca) 82
    - 5.1.2.2 Potassium (K) 82
    - 5.1.2.3 Magnesium (Mg) 83
    - 5.1.2.4 Sodium (Na) 83
  - 5.1.3 TRACE ELEMENTS 84
    - 5.1.3.1 Arsenic (As) 84
    - 5.1.3.2 Barium (Ba) 84
    - 5.1.3.3 Cadmium (Cd) 85
    - 5.1.3.4 Chromium (Cr) 85
    - 5.1.3.5 Copper (Cu) 85
    - 5.1.3.6 Iron (Fe) 86
    - 5.1.3.7 Iodine (I) 86
    - 5.1.3.8 Manganese (Mn) 86
    - 5.1.3.9 Molybdenum (Mo) 87
    - 5.1.3.10 Nickel (Ni) 87
    - 5.1.3.11 Lead (Pb) 87
    - 5.1.3.12 Antimony (Sb) 88
    - 5.1.3.13 Selenium (Se) 88
- 5.2 Soil Sample Analysis Results 89
  - 5.2.1 MAJOR ELEMENTS 89
  - 5.2.2 TRACE ELEMENTS 89

- 5.2.2.1 COBALT 90
- 5.2.2.2 CHROMIUM 90
- 5.2.2.3 COPPER 90
- 5.2.2.4 NICKEL 90
- 5.2.2.5 LEAD 91
- 5.2.2.6 ZINC 91
- 5.3 DISCUSSION OF RESULT 91
  - 5.3.1 MAJOR ELEMENTS IN SOIL AND WATER SAMPLES 91
    - 5.3.1.1 SOIL 91
    - 5.3.1.2 WATER 92
  - 5.3.2 TRACE ELEMENTS IN SOIL AND WATER SAMPLES 92
    - 5.3.2.1 SOIL 92
    - 5.3.2.2 WATER 93
      - 5.3.2.2.1 Arsenic 93
      - 5.3.2.2.2 Selenium 93
      - 5.3.2.2.3 Antimony 94
      - 5.3.2.2.4 Lead 94
      - 5.3.2.2.5 Cadmium 94
  - 5.4 Trace Element Exposure and Human Health 95
    - 5.4.1 Introduction 95
    - 5.4.2 Trace Element Exposure 96
  - 5.5 Trace Elements and Human Health Impact 97

CHAPTER SIX: SUMMARY, CONCLUSION / RECOMMENDATION

- 6.1 SUMMARY 104
- 6.2 CONCLUSION 105
- 6.3 RECOMMENDATIONS 105

REFERENCES CITED   107

APPENDIX   111

CHAPTER ONE: GENERAL INTRODUCTION

1.1 INTRODUCTION

Through physical and chemical weathering processes, rocks break down to form the soils on which the crops that constitute the food supply are raised for humans and animals consumption. Drinking water travels through rocks and soils as part of the hydrological cycle and in the process leached elements in solution (Lar, 2009).

Volcanism and related igneous activities are the principal processes that bring elements to the surface from deep inside the Earth. For example, the volcano Pinatubo ejected on the 2nd of June 1991, about 10 billion tonnes of magma and 20 million tonnes of SO_2 and the resulting aerosols influenced the global climate for 3 years (Selinus, 2004). This event alone introduced 800,000 tonnes of zinc, 600,000 tonnes of copper, and 1,000 tonnes of cadmium to the surface environment. In addition to this, 30,000 tonnes of nickel, 550,000 tonnes of chromium, and 800 tonnes of mercury were also added to the Earth's surface environment. Volcanic eruptions redistribute some of the harmful elements, such as arsenic, beryllium, cadmium, mercury, lead, radon, and uranium. It is also important to realize that there is an average of 60 sub aerial volcanoes erupting on the surface of the Earth at any given time, releasing various elements into the environment. Submarine volcanism is even more significant than that at continental margins, and it has been conservatively estimated that there are at least 3,000 vent fields on the mid ocean ridges (Selinus, 2004).

Almost all metals present in the environment have been biogeochemically cycled since the formation of the Earth. Human activity has introduced additional processes that have increased the rate of redistribution of metals between environmental compartments, particularly since the industrial revolution. However, over most of the Earth's land surface the primary control on the distribution of metals is the geochemistry of the underlying local rocks. Fundamental links between chemistry and mineralogy lead to characteristic geochemical signatures for different rock types. As rocks erode and weather to form soils and sediments, chemistry and mineralogy again influence how much metal remains close to the source, how much is translocated greater distances, and how much is transported in solutions that replenish ground and surface water supplies. In addition, direct processes such as the escape of gases and fluids along major fractures in the Earth's crust, and volcanic related activity, locally can provide significant sources of metals to surface environments, including the atmosphere and sea floor. As a result of these processes the Earth's surface is geochemically inhomogeneous. Regional scale processes lead to large areas with enhanced or depressed metal levels that can cause biological effects due to either toxicity or deficiency if the metals are, or are not, transformed to bioavailable chemical species (Selinus, 2004).

Many elements are essential to plant, human and animal health, but this depends on the dose. Most of these elements are taken into the human body via food, water, in the diet and in the air we breathe.

The naturally occurring elements are not distributed evenly across the surface of the Earth, and problems can arise when element abundances are too low (deficiency) or too high (toxicity). The inability of the environment to provide the correct chemical balance can lead to serious health problems. Approximately 25 of the naturally occurring elements are known to be essential to plant and animal life in trace amounts, including Ca, Mg, Fe, Co, Cu, Zn, P, N, S, Se, I, and Mo. On the other hand, an excess of these elements can cause toxicity problems. Some elements such as As, Cd, Pb, Hg, and Al have no or limited biological function and are generally toxic to humans (Selinus, 2007).

Those living on lands with heavily impoverished soils, have such a low intake of essential elements that a very large percentage of the population suffers from a variety of diseases caused by severe mineral imbalances. Likewise, in areas, where there is excess intake of elements due to the abundance of certain minerals in the environment, may leads to high incidences of toxicity.

Environmental pollution arising from the distribution elements by natural or anthropogenic processes distorts geochemical systems. The natural geochemical composition of rocks and soils that make up the environment where we live may become direct risks to human health and may be the underlying cause of element deficiency and toxicity (lar, 2008).

Because of the increasing concern on the negative effects of excess or lack of trace elements to Humans and Animals an attempt will be made to study trace elements concentration in the soils, surface and underground waters of some part of Biu volcanic province.

1.2 LOCATION, EXTENT AND ACCESSIBILTY

The study area covers some parts of the Biu Plateau. The area is located in the standard sheet 133SW. Lying between longitude 12°07'E and 12°15'E and latitude 10°31'N and 10°38'N. Biu Town is located at the centre of the Plateau. The towns bordering the area include Damaturu to the North, Mubi to the South and Damboa to the East and Gombe to the West figure1.

The area is fairly accessible and has relatively good network of roads and foot paths. There is a trunk 'A' road in the area that stretches from Biu-Damboa road and Biu-Garkida road that give good access for sample collection.

1.3 RELIEF AND DRAINAGE

Topographically, the Biu Plateau stands at an altitude of about 600-800m above sea level, forming a flat top in some areas, it slopes gradually to the north and has steep precipitous escarpment to the south. To the west and east it has steep slopes. There are however gently undulating plains of the buried basement and cretaceous rocks particularly in the western and southern part of the Plateau. The Basement rocks are often deeply weathered, and where the protecting basalt cover has been removed, gullies and rough topography often develop (Du Preeze, 1949).

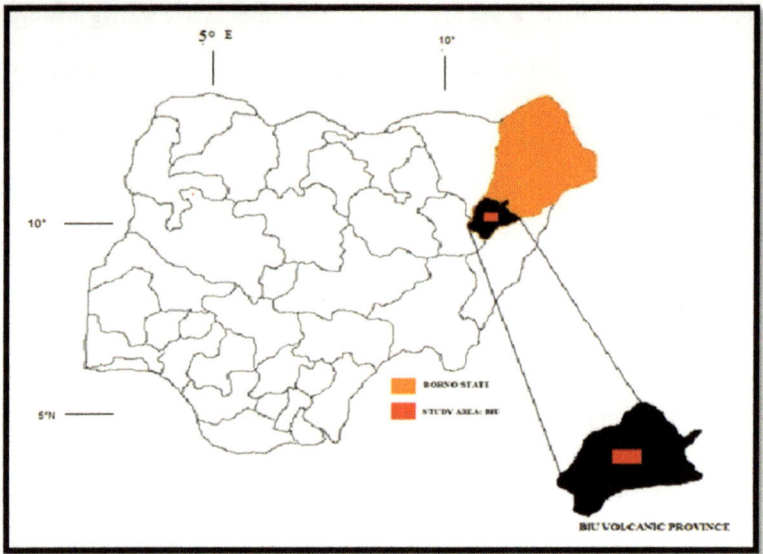

Figure 1: Map of Nigeria showing the location of the study area (Modified after Falconer, 1911)

In the northern part around Miringa area it is characterized by the presence of volcanic cones, which formed many flat top hills. The topography can be observed from the extinct volcanoes in line from north to south along side with Biu-Damboa road, and to the west and east near Zagu.

In the southern part of the area from Kinging to Marama and Lokoja the topography steadily increases and decreases at some interval but generally maintains an average altitude of about 450-600m above sea level. Many hills here have well developed craters with branched rims and steep sided feature. A number of volcanic cones rise above the plain; some appear well preserved but are deeply weathered.

At the south-eastern parts of the Plateau near Kwajaffa there is a sudden drop in the altitude to about 150m above sea level.

Numerous tributaries of the Gongola River including the Hawal, Ruhu, Gungeru, and Ndivana rivers rise on the plateau and deeply dissect its surface. All rivers in the study area are seasonal, displaying dendritic drainage patterns that are both structurally and morphologically controlled (figure 2). While Biu's southern and western sides are quite steep, the plateau slopes more gradually in the north onto the Bauchi Plains and the Chad Basin (Du Preeze, 1949).

1.4 CLIMATE AND VEGETATION

Biu Plateau falls within the Guinea Savannah climatic zones of Nigeria. There are two types of seasons in this area. They are wet and dry season. The wet season starts around April to September while the dry season sets in from September to March (Falconer, 1911).

The month of October to February witness the cold period with extensive cold and dusty cloud. The dry season is influenced by the tropical continental NE trade wind (harmattan), while the wet season with its torrential rains occasionally accompanied by hail storm is induced by the tropical maritime SW trade wind. The plateau receives approximately 1,000 mm (40 in) of precipitation from April to September, the rainy season lasts 140days (Falconer, 1911).

The vegetation of the study area could be best described as Sudan type. It is characterised by trees of about 6- 8 meters high interspersed with tall trees and plain grasses. Vegetation is thicker along the river channels and streams.

Temperature alters very much as it attains about $34°C$ during the day while in the night it could drop significantly to $8°C$.

Relative humidity is generally low, ranging from as low as 13 per cent in the driest months of January and February to the highest values of seventy to eighty per cent in the rainy season months of May and September (Falconer, 1911).

1.5 SETTLEMENT AND LAND USE

The Biu Plateau's thin soils, scarcity of water in the dry season, and relative inaccessibility have discouraged human settlement there. Several ethnic groups (Babur, Bura, Tera, Margi, Hina, and Fulani), make their home on the Biu Plateau, the largest being the Bura people. The Biu plateau has a generally low population density, except for urban pockets in the south. Biu, a regional administrative and commercial centre, is the largest town. Roads link Biu to north-eastern Nigeria's largest cities Maiduguri, Yola, and Gombe. The Biu plateau provided a place of refuge for small, militarily weak groups, who resisted the expansion of the powerful Fulani Sokoto Caliphate, which controlled most of northern Nigeria in the 19th century (Falconer, 1911).

The volcanic soils are naturally fertile but also thin and stony. Small, hand-tilled farms are the mainstay of the region's economy. Sorghum, millet, maize (corn), beans, cotton, and peanuts are the main crops. Some farmers grow crops in terraced plots on the slopes of valleys. Rice is cultivated in some valley bottoms.

Most inhabitants in the region keep cattle, goats, sheep, horses, and donkeys; and Biu town is the chief trade centre (sorghum, millet, peanuts [groundnuts]) on the Plateau. The town, site of the emir's palace, has several (government Health offices and a Dispensary. The Church of the Brethren operates a teacher-training college at nearby Waka Biu (Falconer, 1911).

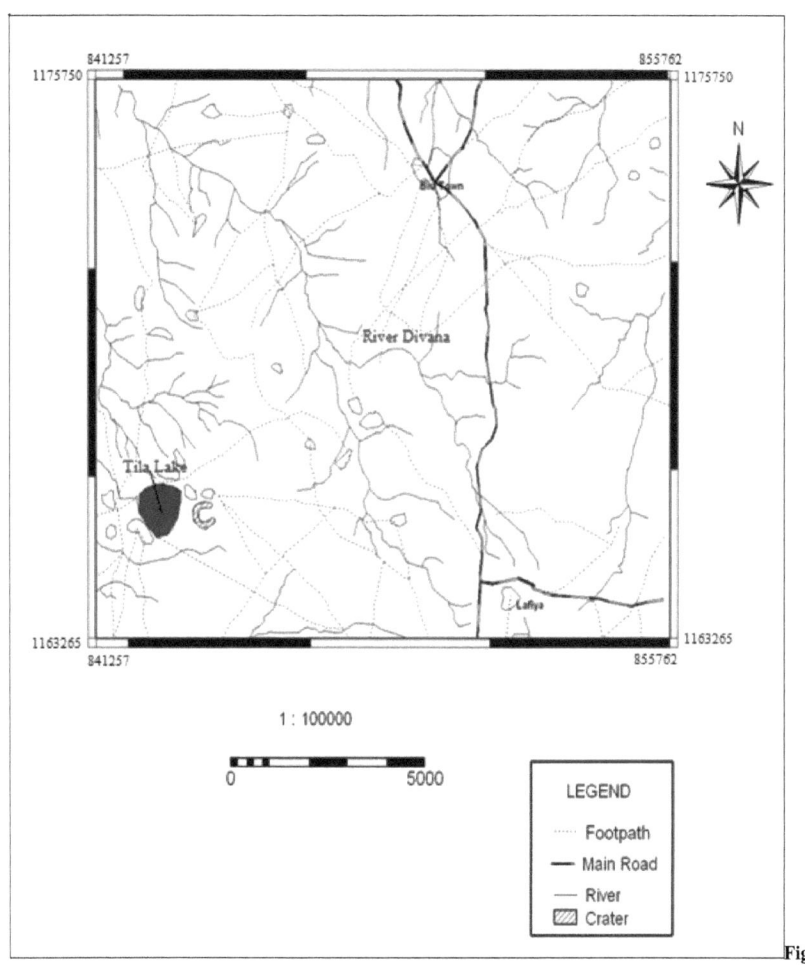

Figure 2: Drainage map of the study area (present study)

CHAPTER TWO: LITERATURE REVIEW

2.1 INTRODUCTION

Not much work has been done on trace element Hydrogeochemistry in volcanic environment in Nigeria. The present study is the first of its kind and will relate the distribution of trace elements from soils into surface and ground waters in the study area to the possible health impact in humans. However, the effects of trace elements in volcanic environment on Human, Animals and Plants have been studied considerably around the world by various authors.

2.2 EFFECTS OF TRACE ELEMENTS IN VOLCANIC AREAS

In the western United States, groundwater with elevated arsenic concentrations is known to be associated with intermediate to felsic volcanic rocks and associated sediments (Welch et al., 1988). Volcanic rocks and sediments derived from them are also associated with elevated arsenic levels in ground water. This may be true not because volcanic rocks contain more arsenic than other types of rocks, but because the arsenic is more readily mobilized from volcanic rocks and derived sediments.

Volcanic rocks and derived sediments in the north-eastern half of the Tucson basin, where sediments are derived largely from granitic rocks, arsenic concentrations in ground water are generally less than 2 parts per billion (ppb). In contrast, in the south-western half of the basin, underlain by sediments derived substantially from volcanic rocks, arsenic is found in much higher concentrations in groundwater (Spencer, 2000). Volcanic rocks and derived sediments, especially where altered to clay minerals, may be especially prone to yielding arsenic to groundwater under such conditions.

The problem of fluorosis related to volcanic activity was first recognised in Japan were this pathology was called "Aso volcano disease" (Kawahara, 1971) due to the fact that fluorosis was widespread in the population living at the foot of this volcano. Water intake being the main route of fluorine into the human body, fluorosis in volcanic areas is generally associated to elevated fluoride content in surface- and ground-waters.

High fluoride concentrations (greater than the WHO guideline value of 1.5 mg/l) are found in the Rift Valley of western Uganda and in the volcanic areas of the east (Mbale, Elgon, Moroto areas). The incidence of fluorosis is known to be high as a result. The crater lakes of western Uganda often have high concentrations (e.g.4.5 mg/l F in Lake Kikorongo; Mungoma, 1990) and concentrations in ground waters having interaction with these lake waters are likewise expected to be high (and the waters correspondingly saline). High fluoride concentrations are particularly noted in groundwaters from the Rwenzori Mountains on the western border and the Sukulu Hills in eastern Uganda (WRAP, 1999). In the Sukulu Hills, fluoride may also be associated with occurrences of phosphate minerals which are currently being investigated for mining development.

Acute and chronic fluorosis on grazing animals has been described for many explosive eruptions all around the world [Mt. Hekla – Iceland (Georgsson and Petursson 1972), Lonquimay – Chile (Araya et al, 1990), Nyamuragira – Democratic Republic of Congo (Casadevall, 1995), Mt. Ruapehu – New Zealand (Cronin et al, 2002). Consequences on livestock are due to grazing of grass or drinking water that is Fluoride -contaminated.

High fluorine content in waters derive either from Water-Rock Interaction (WRI) processes in volcanic aquifers (ground waters) or to contamination due to wet or dry deposition of magmatic fluorine (surface waters - reservoirs). Furthermore, paleopatho-logic studies on human skeletons found in Herculaneum, referable to victims of 79 AD eruption of Mt. Vesuvius, evidenced that fluorosis in this area had the same incidence as in modern times, pointing to the constancy of the geochemical processes responsible for fluorine enrichment of the drinking water in the area over at least the last 2000 years (Morettini and Ciranni, 2000).

Dental fluorosis due to groundwaters enriched by water rock interaction (WRI) in recent or active volcanic areas has been assessed in many parts of the world. Many articles illustrate such cases. Some of them refer to limited areas like Gölcük – SW Turkey (Pekdeger et al, 1992), Mt. Aso volcano, Japan (Kawahara et al, 1971), Island of Tenerife – Spain (Hardisson et al, 2001), Furnas volcano, São Miguel – Azores, Portugal (Baxter et al, 1999), while other evidence a widespread problem throughout entire countries like Mexico (Soto-Rojas et al, 2004), Ethiopia (Kloos and Haimanot, 1999), Kenya (Nyaora et al, 2002), Tanzania (Nanyaro et al, 1984). In these areas, populations as high as 200,000 people could be at risk to develop fluorosis like for example the inhabitants of the Los Altos the Jalisco region in Mexico (Hurtado et al, 2000).

Very acidic lakes in active volcanic systems can also achieve extreme fluorine concentrations not only due to intense WRI processes but also to direct input of F-rich volcanic gases. Lake like Ijen Crater Lake – Indonesia (Heikens et al, 2005) reach concentrations far above 1000 mg/l. Seepage or effluent rivers from these extremely F-rich lakes can easily contaminate ground- or surface waters. It has been estimated that the Ijen Crater Lake discharges daily in the surface and ground waters of the highly populated area of Asembagus about 2800 kg of fluorine (Heikens et al, 2005), which is responsible of the widespread occurrence of fluorosis in the area. Furthermore the fluorine contained in the salts extracted from the shores of the East African Rift Valley lakes and used for cooking purposes represent an additional fluorine source for the local population (Nanyaro et al, 1984).

Ullrey (1981) reported the occurrence of low selenium areas in Arizona and New Mexico where the soil is formed from Tertiary volcanic rock. These areas also tend to correspond to areas where selenium-deficiency disorders such as white muscle disease occur.

Volcanic eruptions that discharge more than a few mega tonnes of sulphur gases into the atmosphere have the potential to cause regional to global scale climate change through complex mechanisms. A well observed and understood phenomenon is the summer cooling and winter warming of Northern Hemisphere continental regions following large, sulphur-rich eruptions of volcanoes in the tropics. The 1815 eruption of Tambora, Sumbawa Island, Indonesia, responsible for the greatest recorded fatalities due to volcanic activity (Witham, 2005), also released sufficient sulphur into the upper atmosphere to result in widespread cooling during the Northern Hemisphere summer in 1816. This has been linked to epidemic disease in Ireland, the UK, and parts of continental Europe through a combination of socioeconomic factors and the effects of the climatological anomalies on crop yields (Oppenheimer, 2003). In the immediately impacted region, an estimated 61,000 people died during and in the aftermath of the eruption, mostly as a result of famine and epidemic disease.

CHAPTER THREE: DETAILED GEOLOGY AND HYDROGEOLOGY OF THE STUDY AREA

3.1 DETAILED GEOLOGY

3.1.1 INTRODUCTION

Biu Plateau is situated on the structural and topographic divide between the Benue and Chad sedimentary basins (fig. 3). The structural divide is a broad E-W ridge or swell of basement, which extends to the western edge of the Biu Plateau. The two basins are divided by the Zambuk ridge to the west (Carter et al., 1963). Fig 4 shows a simplified geological map of Nigeria.

The basalt of the Biu Plateau mainly overlies Basement rocks. These are predominantly granites, granite-gneiss and Fayalite-quartz, Monzonite, Bauchites (near Wandali at the SW margin of the plateau), hypersthenes diorite, volcanic and sub volcanic rocks of the Burashika group (Turner, 1978). To the west and north, Basalt of the Biu Plateau has spread over cretaceous sediments, mainly the arkosic Bima sandstone. These rocks are folded, with axes to the SW of the Plateau trending NE-SW, the structures extending into the basement rocks as NE-SW faults (Turner, 1978).

The buried landscape consisted of gently undulating plains on both the basement and Cretaceous rocks, which must have been particularly featureless in the NE. (Du-Preez, 1949) discovered that the sub-basalt surface was lateralized. Laterite exposed beneath the basalt near Puba, 4km SW of Kwajaffa consists of about 1.5m of hard pisolitic ironstone overlying residual clays and weathered granite-gneiss.

Two important post depositional processes affect the Poliocene basalts. The first was internal, the crystallization of zeolites and calcite, which are abundant in vesicular and rubbly interflow horizons. The second is surface weathering to clays and laterites. The basalt of the poorly drained northeastern plains is deeply decomposed to clay presumably a continuing process. Much less widespread, but more significant, is the development of laterite on the high Plateau.

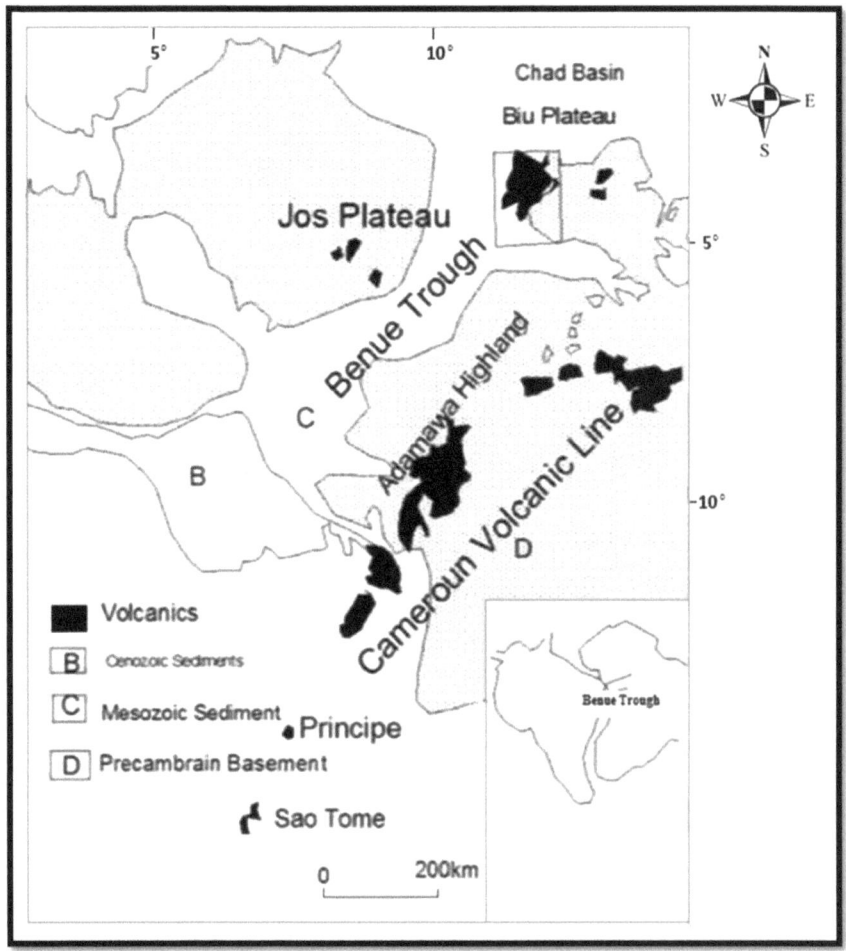

Fig.3: Location MAP of Biu Plateau and other Rock types in Nigeria (adopted from Wright, 1976)

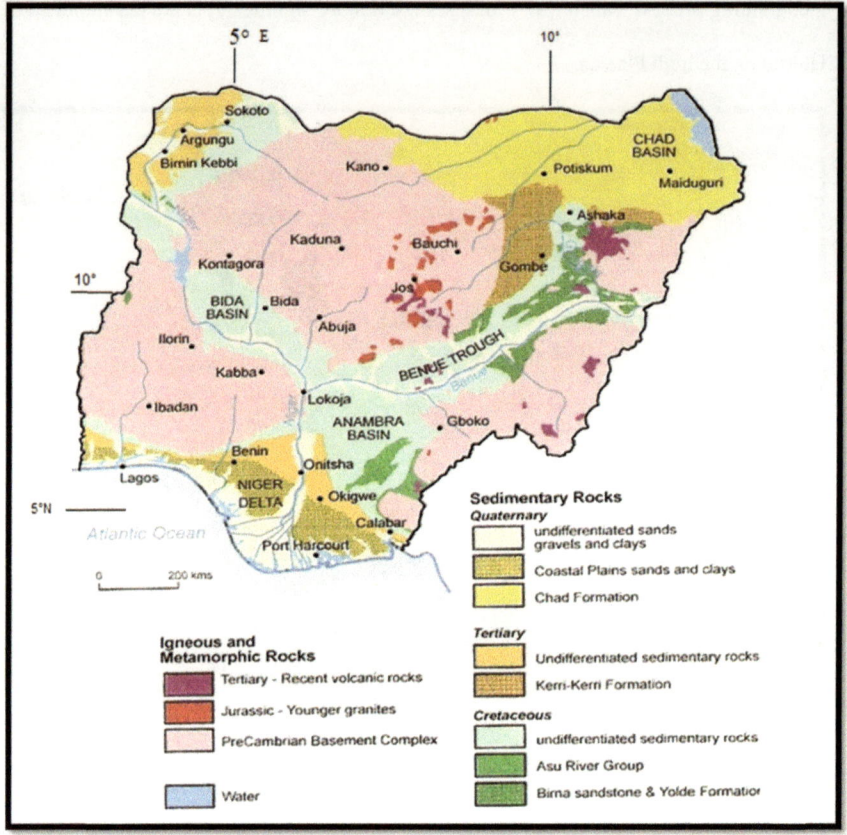

Figure 4: A Simplified Geological Map of Nigeria (Source: - NGSA, 2004)

The rocks in the Biu Plateau mostly occur as "flood basalts" in a number of flows and in fact cover nearly 85% of the area with its centre around Biu (Turner, 1978). The basalt at some places has built up large number of flows. The dimension of the flows and the marked absence of pyroclastics in and around Biu, Tum, Marama, and Shaffa areas, indicate that the eruption of basaltic magma in these places was not violent. However, the basaltic sequence in the Northwestern part of Biu (Miringa area, fig. 5) is surrounded by several youthful scoria, cinder cones, tephra rings etc, the pyroclastics are generally restricted to the area west of Biu- Damaturu road, suggesting that the eruptions in these places are violent in nature.

The volcanoes are built up by essentially basaltic materials and are of two textural types namely: the Biu type, which is flow basalts, massive with vesicles, and the Maringa type, highly scoraceous and associated with pyroclastic deposits. These two types present the same mineralogy consisting of phenocrysts of both olivine, plagioclase (bytownite-labradorite), and rarely pyroxene (diopside-augite) set in a groundmass of labradorite laths, magnetite, ilmenite, minor K-feldspars, nepheline and volcanic Glass (Saidu, 2004).

The Basement rocks on which the basalt of the Biu Plateau overlies are predominantly granites and granite-gneisses. The foot of the Basement can be seen in the southern part of Biu (Kwajaffa area), Southwestern part (Wandali area) and Southeastern part (Garkida area). To the North and Western part of the Biu Plateau the basalt spread over Cretaceous sediments, mainly over the arkosic Bima sandstone (Turner, 1978).

Du Preez, (1949), studied the detailed petrology of the Biu basalt; petrologically the basalts are represented by a wide range of rock types such as porphyritic and amygdaloidal variety, large phenocryst of zoned labradorite and agglomerates, tuffs and brownish red basaltic scoriae. The land sat image (fig 6) gives a synoptic view of the study area.

3.1.2 PORPHYRITIC AND AMYGDALOIDAL VARIETY

They occur in the central Plateau around Biu showing the lath of labradorite feldspar, some of which exhibit a rudimentary trachytic arrangement. Large phenocrysts of olivine are common but smaller crystals also constitute an important proportion of the groundmass, in some cases the smaller crystals are completely altered to iddingsite, whereas the phenocrysts are either unchanged or show only marginal alteration effects. Some amygdales are encrusted or completely filled with zeolites such as stilbite, and sometimes with calcite. Pyroxene is present as small elongated crystals in the groundmass and iron oxide is abundant.

3.1.3 LARGE PHENOCRYSTS OF ZONED LABRADORITE

They formed a prominent constituent of the worn volcanic cones near Jaragwol on the Gongola. The groundmass, which consist chiefly of labradorite laths, minutes grain of pyroxene and much iron ore, shows pronounced flow structure, especially around the large feldspar and olivine phenocrysts. The olivine shows extensive serpentinisation. The basalts at Ngulde exhibit a glomero-phorphyritic texture.

The phenocrysts of labradorite and mafic minerals are arranged in clusters, the former usually forming an outer rim enclosing a core of ferro-magnesium minerals consisting chiefly of enstatite. The groundmass sometimes shows an ophitic texture and consists of laths of labradorite together with olivine, clino-pyroxene and some enstatite. Large beautifully zoned labradorite phenocrysts are often present and enstatite is sometimes seen to change into clini-pyronexe.

3.1.4 AGGLOMERATES, TUFFS AND BROWNISH RED BASALTIC SCORIA

They are widely distributed in Babur district. Scoriaceous are not infrequently found inter-bedded with black, fine-grained varieties. The volcanic cones to the west of Biu-Damaturu road are composed of decomposed scoriae, fragments of ropy lava, bombs, tuffs and agglomerate, the latter frequently containing boulders of granite-gneiss

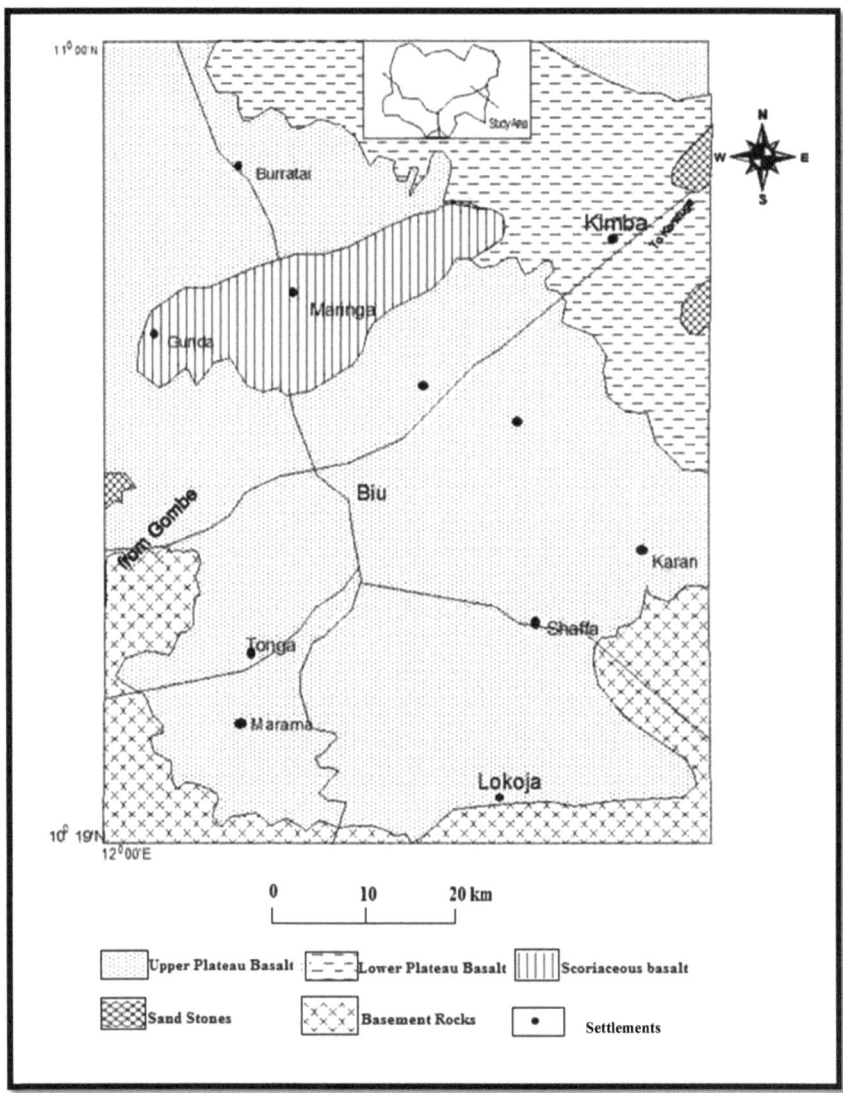

Fig.5: Geological map on Biu-plateau (adopted from Saidu, 2004)

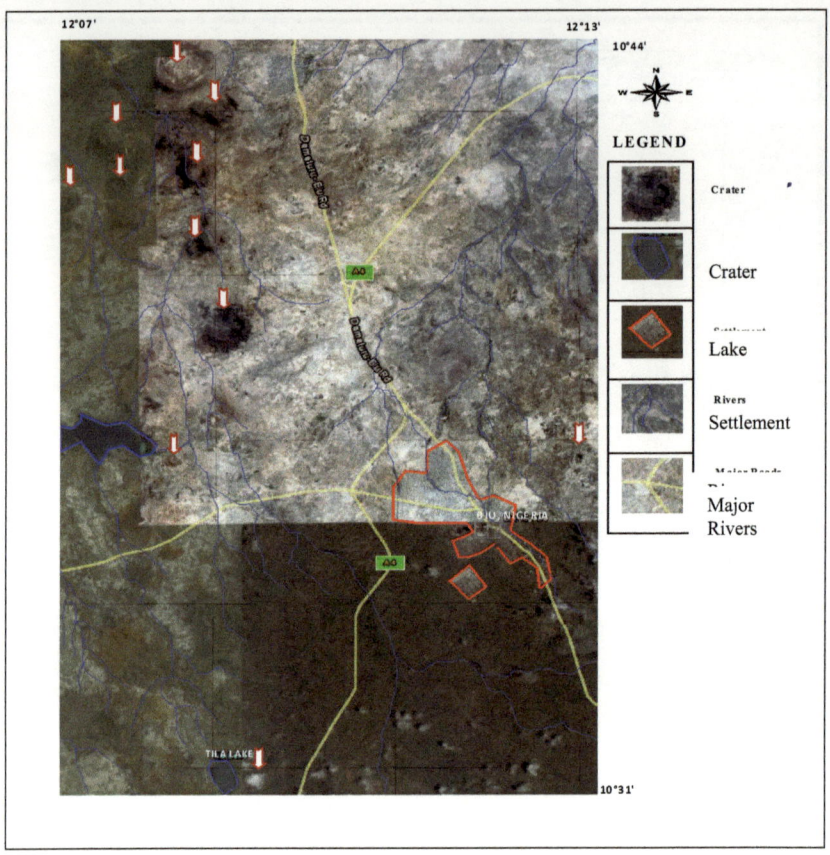

Figure 6: Landsat image of the Study Area, insert are craters and Tila Lake (from Google Maps, 2011)

Excellent exposures of bedded tuff dipping at a low angle to the north-west may be seen at Lake Tila. Agglomeratic tuffs form aprons around the prominent volcanic cones one mile north of Miringa. In this locality, masses of agglomerate consisting chiefly of large and small lava bombs, inclusion of ultra-basic rocks and granite-gneiss build picturesque cliffs and crags in which caves have been formed in some places. Similar outcrops are present in many other localities notably at Mount Sagu north of Buratai. Mount Tila is composed of brownish-red Scoriaceous basalts. South of Biu agglomerate are not abundant, there are however, several outcrops of agglomerate near Ngwa and also an interesting occurrence at a much lower level several miles west of Buma on the Hawal.

Isolated occurrences, like Mount Biski three miles north-west of Haran in East-Bura district, are sometimes found.

Ultrabasic nodules are an important component of the agglomerates in some areas. West of Miringa they may be up to one foot in diameter. Frequently they form the cores of lava bombs with an outer shell of lava froth. They generally consist essentially of olivine, but diopside is sometimes present. Another interesting type of inclusion observed west of Miringa consists exclusively of pyrope with a network of kelyphitic veinlets. Red grains of pyrope have also been found in the heavy mineral concentrates in the Magzar River some eight miles east of Miringa. One inclusion was seen to consist of pyrope with subordinate olivine in diopside showing definite kimberlitic affinities. Iddingsite nodules were seen in some localities, but in general they are rare. Other bombs containing large crystals of glassy oligoclase and dark green olivine have been noted.

3.2 HYDROGEOLOGY OF THE STUDY AREA

3.2.1 INTRODUCTION

Water is the most important natural resource provided for survival of man by nature. Globally, water is mostly used for domestic, industrial and agricultural purposes. The water resources of the study area can be divided into surface and groundwater resources.

The surface water of this area occurs in the form of streams and lakes. They serve as water supply sources for both drinking and domestic uses. Most of the streams are seasonal. The streams and lakes are recharged by direct precipitation during the rainy season.

3.2.2 SURFACE WATER

3.2.1.1 LAKE TILA

The only permanent body of water is Lake Tila situated at Kwaya Bura village about thirteen kilometres to the south-west of Biu. The lake occupies a crater a short distance west of Mount Tila, Plate 1.

The lake-level in the height of the dry season is only one meter below what appears to be the peak level bench-mark. The high water level each year is dependent on the rainfall and it can be assumed that it ranges within one meter, (Du Preez, 1965).

The lake receives very little runoff during the rains and the rise of the lake level is due mainly to direct precipitation. There is little seepage zone around the shore which flows continuously into the lake. Shallow percolation wells dug into the zone constitute the main domestic water supply for several villages in the vicinity.

Analyses of water samples from Lake Tila and from the shallow seepage wells along the shoreline carried out by the (Du Preez, 1965) are given in Table1. The analysis of the lake water shows a high concentration of sodium carbonate, and comparatively low sodium chloride content. In the water from the seepage wells, sodium carbonate probably also predominate, while sodium chloride is negligible. Owing of the high carbonate content the lake water is unsuitable for irrigation. It is estimated that the amount of soda in the lake is in the order of about 2,000 tons, but it is doubtful whether it is of any economic value, (Du Preez, 1965).

Plate 1: Tila Lake

Table 1: Analysis of water sample from Lake Tila (Source: Du Preeze, 1965)

Source	Cl (mg/l)	Fe (mg/l)	Total Alkalinity as $CaCO_3$ (mg/l)	Total Solids (mg/l)	After gentle ignition (mg/l)	pH
Seepage wells	2.0	0.08	410.0	500.0	346.0	7.6
Lake Tila	87.0	nil	2403.0	2846.0	2572.0	9.5

3.2.1.2 PERENNIAL STREAM WATER

Most of the rivers in Biu dry up for a part of the year, but a considerable number of streams rising on the basalt Plateau are weakly perennial some of the streams are fed by seepage springs which issues from the basalts, (plate 2). These discharges are small and are absorbed in the alluvium on reaching the low-lying country below the basalt Plateau. Permanent pools are frequently present in these streams, and they provide water supplies for neighbouring villages.

3.2.1.3 GROUNDWATER IN THE BASALTS

The basalts form the most important aquifer in Biu Plateau. The amount of water obtained in them depends on the degree of decomposition, jointing and the nature of the rock. The basalts show a superficial weathered zone consisting of black-grey and brown clay with residual boulders of partially decomposed rock.

The scoriaceous and amygdaloidal basalts usually weather to brown and blue clays which tend to disintegrate on exposure to the atmosphere. Moderate amounts of water may be present in the basalt alluvium (Plate 3a and 3b) but in general the quantity of water available is small and the water table is subject to considerable variation. Many villagers get their water partly or wholly from this source during the dry season. The water from this source can be obtained through hand dug wells which were seen all over the area and few boreholes.

Plate 2: Water Spring in basalts at Yimirshika (Longitude 12° 14'41.4" and Latitude 10° 31'52.2", Elevation 850m)

Plate 3a: Woman fetching water from a water spot in the basalt alluvium (Longitude 12°07'57.7", Latitude 10°35'35.4", Elevation 659m)

Plate 3b: Moderate amount of water in shallow well (basalt alluvium) Longitude 12°09'55.6" and Latitude 10°36'05.4"

The basalts are usually strongly jointed and fissured. The earlier flows, usually consist of dense compact basalt, while the early flows are irregular jointed. The joints serve as loci of weathering and as channels for the circulation and storage of groundwater (Du Preez, 1965). The compact basalts are incapable of storing water, the groundwater held in the joints gives rise to numerous small springs on the higher parts of the Plateau.

3.2.3 DATA COLLECTION

Data of the water table was obtained by field measurement of the static water levels in wells within the study area, as well as the geographical positions and elevations at various locations Table 2.

3.2.4 DATA PROCESSING

Coordinates of the observed wells and well elevations were manipulated using a computer program (sufer 8). The X, Y, Z coordinates in the surfer 8 is entered as longitude, latitude and elevation respectively, from which the computer program plots a water table contour map (figure 7).

3.2.5 INTERPRETATION OF GROUNDWATER MAP

A water table contour map shows the elevation and the configuration of the water table at a certain data. The map is prepared by plotting the absolute water levels of all observation points of equal water table elevation. This water table contour map (figure 7) is an important tool in groundwater investigations as one can derive from it the gradient of the water table and the direction of the groundwater flow.

Generally, the pattern of groundwater flow follows the surface topography with seasonal variations in water levels characterized by rising water levels during the wet months and declining water levels during the dry months. In the study area, the dominant groundwater flow direction is NW (figure 7).

Table 2: Static water level and hydraulic gradient of the study area

Sample(ID)	Source	Locality	Longitude	Latitude	Elevations(m)	SWL	Hydraulic Head
BH1	Borehole	Waka	12°10'50.9"	10°38'00.3"	688	Nil	Nil
BH2	Borehole	Hema	12°09'48.1"	10°34'02.5"	750	Nil	Nil
BH3	Borehole	Barrack	12°11'46.4"	10°35'49.5"	775	Nil	Nil
BH4	Borehole	Biu	12°11'38.7"	10°36'17.4"	767	Nil	Nil
W1	Well	BCG	12°07'40.2"	10°36'48.4"10"	650	7	643
W2	Well	Wakama	12°07'41.5"	36'5vr27.3"	643	8	635
W3	Well	Waka	12°14'38.5"	10°31'55.8"	829	7	822
W4	Well	Waka	12°13'07"	10°31'17.5"	837	4	833
W5	Well	Biu	12°12'40.3"	10°36'40.5"	711	2	709
W6	Well	Biladega	12°12'47"	10°34'25.4"	770	4	767
W7	Well	Tabra Fulani	12°07'57.7"	10°35'35.4"	659	4	655
W8	Well	Malan	12°08'22.8"	10°35'35.4"	681	4	677
W9	Well	Takwa	12°08'37.8"	10°33'31.4"	736	2	734
W10	Well	Tan	12°08'05.1"	10°35'51.6"	722	7	715
W11	Well	Tila	12°09'18.2"	10°33'13.3"	775	23	752
W12	Well	Yimishika	12°10'15.6"	10°37'11.5"	742	11	731
W13	Well	Tila	12°09'46.6"	10°34'37.4"	723	7	716
W14	Well	BCG	12°09'47.3"	10°34'11.5"	727	3	724
W15	Well	BCG	12°12'04.2"	10°36'20"	772	6	766
W16	Well	Yimishika	12°12'29.8"	10°35'14.8"	793	5	788
W17	Well	Gwarta	12°12'49.8"	10°35'	792	8	784
W18	Well	Biu	12°12'27.6"	10°32'38.1"	759	3	756
W19	Well	Filin Jirgi	12°12'12.9"	10°31'20.1"	822	57	822
W20	Well	Biladega	12°10'31.9"	10°33'36.6"	734	1	733
SW1	Stream	Biladega	12°08'00.9"	10°37'04.6"	850	0	0
SW2	Stream	Tila	12°10'48.4"	10°37'49.7"	711	0	0
SW3	Stream	Tila	12°11'44.1"	10°37'57.8"	670	0	0
SW4	Stream	Tila	12°13'18.2"	10°37'23.1"	725	0	0
SW5	Stream	Tabra Fulani	12°08'03.2"	10°34'26.8"	722	0	0
SW6	Stream	Hena	12°10'32.4"	10°36'20.9"	699	0	0
SW7	Stream	Hena	12°09'55.6"	10°36'05.4"	725	0	0
SW8	Stream	Hena	12°10'48.1"	10°36'01.4"	700	0	0
SW9	Stream	Barrack	12°10'52.9"	10°32'39.1"	747	0	0
SW10	Stream	Barrack	12°10'41.8"	10°33'36.6"	758	0	0
SW11	Stream	Kunar	12°11'39.2"	10°36'55.6"	729	0	0
SPW	Spring	Yimishika	12°14'41.4"	10°31'52.2"	850	0	0
TLK	Lake	Tan	12°07'59.6"	10°31'48.9"	809	0	0

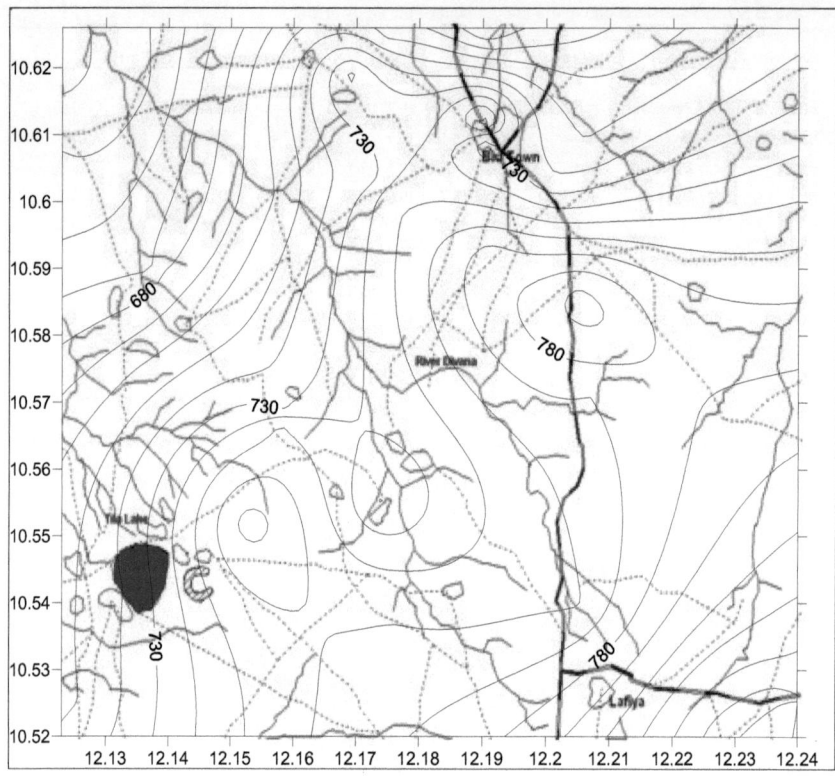
Figure 7: Groundwater Contour Map of the study area indicating direction of Water Flow (present study)

3.2.6 STRUCTURES

Structural features are resultant effects of deformation in rocks. They are produced on rocks as a result of stresses and strains. The variation in deformation both in intensity and area extent depends on the nature of the rocks involved, the intensity and duration of deformation, as well as the number of times such rocks have been deformed coupled with pre-existing structures on the rock.

The structural features that were observed in the study are joints, their attitude and direction is shown in appendix 1. These structures exhibit diverse trends in the basalts and all are in conformity with the general structural trends of North-Eastern Nigeria.

3.2.7 JOINTS

In the study area, joints are not well exposed except in few areas, probably due to thick layers of boulders overlying the basalts. However, the dominant trends of the observed joints in the basalts are NE-SW, NNE-SSW, E-W (figure 8) corresponding to the major structural trends in the Land Sat image of the study area (figure 9).

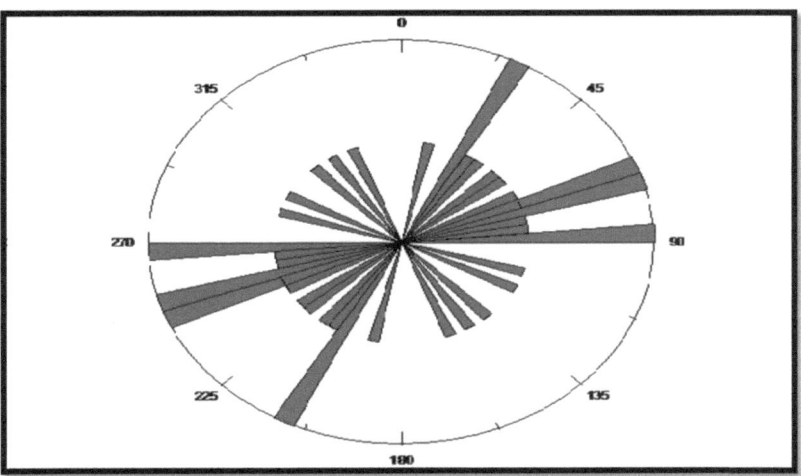

Figure 8: Rose diagram of study area showing the general structural trends NE-SW, NNE-SSW, and E-W of the observed Joints

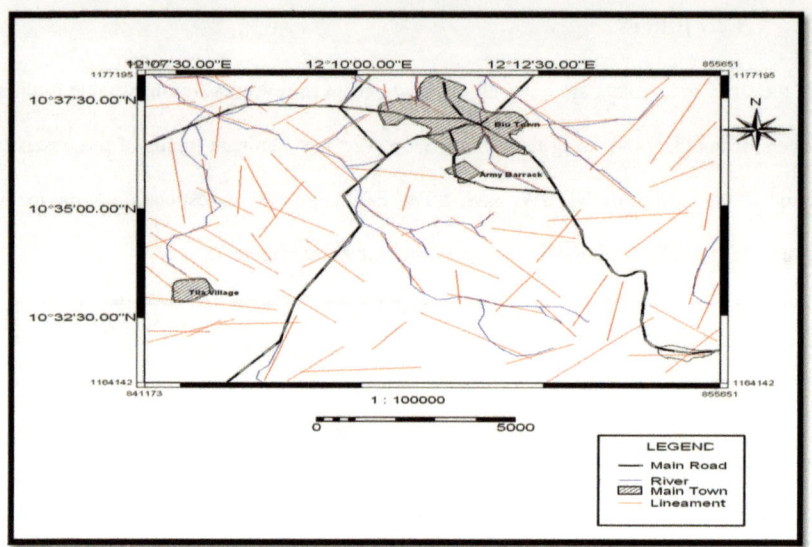

Figure 9: Lineation map of study area showing the general trends NE-SW, NNE-SSW, and E-W directions (present study)

CHAPTER FOUR: HYDROGEOCHEMISTRY

4.1 INTRODUCTION

Groundwater originates from infiltration of rain and surface water. The chemical composition is the combined result of the composition of water that enters the groundwater reservoir and reactions with minerals present in the rock that may modify the water composition; the water takes into solution different concentrations of some elements that make up the rock. The chemical composition could also be influenced by anthropogenic sources. Other factors that controlled the chemistry of groundwater in any geological environment include primarily the chemistry of infiltrating water at the recharge source and the chemistry of the porous medium, including the interstitial cement matrix of the aquifer. Secondly, the rate of groundwater flows in the aquifer and the travel time of the water through the environment. The geochemical sampling media considered in this work are; surface water, groundwater and soils.

4.2 METHODOLOGY

4.2.1 SAMPLE COLLECTION AND PRESERVATION

Sampling was accomplished during the dry season (March 28-April 25, 2009). A total of thirty seven (37) acidified water samples and thirteen (13) soils samples were collected over an area of 150 km^2 for analysis to determine their trace element concentrations.

Table 3 shows water samples obtained from; borehole (4), streams (11), wells (20), spring (1), lake (1), while the soils samples are obtained from farms, Table 4.

Table 3: Water samples description and location

Sample ID No	Source	Locality	Longitude	Latitude
BH1	Borehole	Waka	12°10'50.9"	10°38'00.3"
BH2	Borehole	Hema	12°09'48.1"	10°34'02.5"
BH3	Borehole	Ar. Barrack	12°11'46.4"	10°35'49.5"
BH4	Borehole	Biu	12°11'38.7"	10°36'17.4"
W1	Well	BCG	12°07'40.2"	10°36'48.4"
W2	Well	Wakama	12°07'41.5"	10°36'57.3"
W3	Well	Waka	12°14'38.5"	10°31'55.8"
W4	Well	Waka	12°13'07"	10°31'17.5"
W5	Well	Biu	12°12'40.3"	10°36'40.5"
W6	Well	Biladega	12°12'47"	10°34'25.4"
W7	Well	Tabra Fulani	12°07'57.7"	10°35'35.4"
W8	Well	Malan	12°08'22.8"	10°35'35.4"
W9	Well	Tila	12°08'37.8"	10°33'31.4"
W10	Well	Tila	12°08'05.1"	10°35'51.6"
W11	Well	Tila	12°09'18.2"	10°33'13.3"
W12	Well	Yimishika	12°10'15.6"	10°37'11.5"
W13	Well	Hema	12°09'47.3"	10°34'11.5"
W14	Well	Hema	12°09'46.6"	10°34'37.4"
W15	Well	BCG	12°12'04.2"	10°36'20"
W16	Well	Yimishika	12°12'29.8"	10°35'14.8"
W17	Well	Gwarta	12°12'49.8"	10°35'
W18	Well	Kunar	12°12'27.6"	10°32'38.1"
W19	Well	Filin Jirgi	12°12'12.9"	10°31'20.1"
W20	Well	Biladega	12°10'31.9"	10°33'36.6"
SW1	Stream	Wakama	12°08'00.9"	10°37'04.6"
SW2	Stream	Waka	12°10'48.4"	10°37'49.7"
SW3	Stream	Tila	12°11'44.1"	10°37'57.8"
SW4	Stream	Tila	12°13'18.2"	10°37'23.1"
SW5	Stream	Tabra Fulani	12°08'03.2"	10°34'26.8"
SW6	Stream	Hema	12°10'32.4"	10°36'20.9"
SW7	Stream	Hema	12°09'55.6"	10°36'05.4"
SW8	Stream	Hema	12°10'48.1"	10°36'01.4"
SW9	Stream	Barrack	12°10'52.9"	10°32'39.1"
SW10	Stream	Barrack	12°10'41.8"	10°33'36.6"
SW11	Stream	Kunar	12°11'39.2"	10°36'55.6"
SPW	Spring	Yimishika	12°14'41.4"	10°31'52.2"
TLK	Lake	Tila	12°07'59.6"	10°31'48.9"

Table 4: Soil sample description and locations

Sample ID No	Locality	Longitude	Latitude
AD1	BCG	12°07'06.9"	10°36'21.5"
AD2	Yimishika	12°1'43.9"	10°31'48.2"
AD3	Gwarta	12°13'06.8"	10°31'16.8"
AD4	Waka	12°11'14.9"	10°37'56.9"
AD5	Waka	12°12'50.9"	10°36'51.1"
AD6	Filin Jirgi	12°12'48.1"	10°34'36.4"
AD7	Biladega	12°07'57.9"	10°35'34.7"
AD8	Biladega	12°08'03.2"	10°34'26.8"
AD9	Tila	12°08'01.9"	10°32'50.3"
AD10	Tila	12°08'55.3"	10°32'50.3"
AD11	Malan	12°09'55.6"	10°36'08.1"
AD12	Kunar	12°12'30.2"	10°32'45.2"
AD13	Takwa	12°11'13.7"	10°32'29.8"

It was not possible to follow accurately the initial regular sampling of 1km x 1km because mountainous terrain and irregular distribution of wells, streams and ponds in the study area are dry. The sampling is done at the peak of dry season. However, care was taken to preserve a uniform distribution of sampling sites over the study area. The sampling locations of the study area are presented in Figure (10).

The soil samples collected were stored in clean polythene bags and labelled. The location of the sample is then noted on the base map with the aid of a GPS. The GPS reading was also recorded together with other information about the area in the field note book.

In view of the usual low trace element concentrations in water, various measures were taken to prevent the slightest contamination in the collected samples. 250ml polyethylene bottle capacity containers were used for the collection of samples.

The sample bottles were first washed with a mixture of acid and distilled water (1% HNO_3). The bottles were finally rinsed with distilled water and kept to dry in an oven at 25^0C. One important step taken was the immediate wrapping of the bottle with sterilized thin film with the top of the bottle folding over a non-contaminating stiffened material attached to the twisted end. With these procedures the bottles were protected and ready for sample collection.

At each time and place where a water sample (surface or groundwater) was collected, it was acidified with one or two drops of HNO_3 to keep the ions in solution and to prevent their absorption and precipitation in solution. A test for pH, temperature and conductivity were done using the pH measuring metre. This was achieved in the field by dipping the metre into the sample bottle immediately after collection and the reading is taken after some few minutes when the value is stable. The pH, electrical conductivity and temperature of all the water samples were measured in the field (Table 5).

After the sample was collected, the old thin film was removed and a new one re-wrap. All samples collected were labelled according to location, nature of sample, date of collection and number. Samples collected were kept in the refrigerator at the Department of Geology and Mining in the Geochemistry Laboratory at room temperature (23^0C).

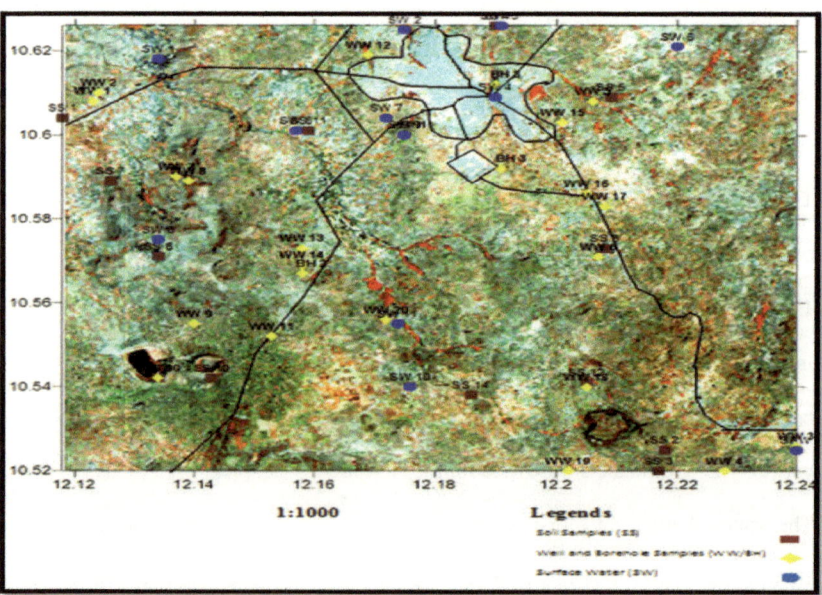

Fig. 10: sampling locations of the study area (this study)

4.2.2 WATER SAMPLE PREPARATIONS

The ICP technique is solution based and therefore water samples will require no special treatment as in the case of solid rocks. The acidified water sample is filtered in a 100ml conical flask and then introduced into the ICPOES machine for analysis.

4.2.3 SOIL SAMPLE PREPARATIONS

The soil samples collected in the field are completely dry therefore, required no further drying in the laboratory.

The procedure of the sample preparations are as follows:

- The soil sample is pulverized with Agate Mortar
- 0.1g of the pulverized sample is weighed in a Teflon crucible
- Aqua regia i.e. (HCL and HNO_3) in the ratio of 6:3 was used to digest the sample
- The digested sample is heated in a sand bath in a fume cupboard
- 2ml of HCL and 10ml of distilled water was added to dissolve the digested sample
- The dissolved digested sample is filtered and made up to 100ml in a volumetric flask
- The sample is passed into ICPOES machine for analysis

4.3 ANALYTICAL TECHNIQUE

Trace and major elements were analyzed using Inductively Coupled Plasma Optical Emission Spectrometry (ICP-OES) at Geochemistry Laboratory, University of Jos (Plate 4). ICP-OES is a multi-element technique, capable of measuring up to 70 elements at very low detection limits (ppm to ppb) in just about any material or substance (waters, biological materials, inorganic materials of all sorts, environmental samples, geological samples, etc.) in solution. Most element of the Periodic Table (with the exceptions of H, O, N, F, Cl, Br, I) can be measured using ICP-OES. The analytical range of solution ICP-MS extends from the ppb (parts per billion) to the ppm (parts per million) regions.

The ICP technique is solution based and therefore water samples will require no special treatment as in the case of solid rocks. The water sample is then introduced into radio frequency excited plasma (~8000°C).

Atoms within the samples are excited to the point that they emit Photons. The number of photons produced is directly related to the concentration of that element in the sample.

Both the accuracy and precision of ICP-OES measurements is dependent, in part, upon the calibration technique used in most cases, accuracies and precisions of ~ 1 - 3 % may be expected. The most common calibration technique options for ICP-OES measurements are calibration curve and standard additions. In addition, the option of using internal standardization is available for the calibration curve technique and the ability of matrix matching. ICP-OES has the added option of using an internal standard that has a known concentration of the element measured.

For quantitative ICP-OES analysis, calibration is most commonly achieved by external standardisation. The signal intensities of all analyte isotopes are measured in a blank as well as in one or more artificial or natural standards with different, known analyte concentrations that cover the concentration range of interest. The (hopefully) linear relationship between the blank-corrected standards on a diagram of signal intensity versus concentration is used to establish a calibration curve that may be used to calculate the concentration of the analytes in samples of unknown composition.

The precision of an ICP measurement is a function of many factors:

- Try to keep the analyte concentration well within the linear working range. When the concentration falls < 100 times the detection limit, the precision begins to fall.
- Avoid lines requiring spectral correction.
- Avoid lines that are in spectrally complex regions requiring sophisticated background correction algorithms.
- Increasing the integration time to as high as 5 seconds should improve the precision. It should not be increased any higher.
- Use an all glass introduction system including a glass concentric nebulizer. Your washout times will improve along with your precision.
- Be wary of salting out effects when analyzing high salt containing solutions, but do not use high solids nebulizers unless necessary.
- If your samples and standards have the same background then you should skip background correction.
- If the matrices of the samples and standards can be matched then eliminate the peristaltic pump and go 'free flow'. The pump introduces pulsing and the pump tubing stretches, causing the introduction rate to gradually change. If you can avoid the pump then make sure the liquid levels of your samples and standards are the same.

NOTE: Slight differences in hydrostatic pressure will make a difference.

- Do not begin the measurement unless the instrument has been allowed to warm up for at least an hour. Additionally, working in a temperature-controlled atmosphere is very advantageous.

- Clean your introduction system and replace the tubing when a) you notice an increase in precision when using glass introduction systems, b) you notice a 'bend or crink' in the Teflon tubing, c) the torch begins to build up residue.

A total of 25 elements were analyzed (Ba, As, Ca, Cd, Co, Cr, Cu, Fe, I, K, Mg, Mn, Mo, Na, Ni, P, Pb, S, Sb, Sc, Se, Sr, Ti, V, Z) for surface and ground water and 20 elements (Mg, K, Ca, Ti, V, Cr, Mn, Fe, Co, Ni, Cu, Zn, As, Se, Mo, Cd, Sb, Pb, Be, Tl) for soil samples.

Plate 4: Photograph of Sample of ICP OES machine in Geochemistry Laboratory University of Jos

CHAPTER FIVE: PRESENTATION OF RESULTS/DISCUSSIONS

5.1 PRESENTATION OF RESULTS

The result of the analysis for major and trace element concentration of surface and ground waters and measured pH are presented in Table 5, 6a, 6b and 7.

Comparison between trace elements and World Health Organization (WHO, 2008) admissible levels for drinking waters are presented in Table 8. While Average Chemical Composition of Parent Rock (ACCPR) of Biu Plateau Basalt (Saidu, 2004), (table 9) will be used as reference material to understand the distribution and concentration of the elements in volcanic soils of the study area. The average concentration of major and trace elements in Soils, surface water and groundwater are compared in Table 10

Data obtained were processed using Ilwis, Surfer, Grapher and Microsoft Excel software's were used to process the raw data. The data are presented as contours, bar chats and 3D Surface Maps as shown in figures (11-39), which are useful tools in summarizing the salient facts drawn from the analysis.

5.1.1 Water Sample

Because of worldwide awareness towards water pollution problems, drinking water standards were developed based on toxicity levels of numerous substances, which are commonly detected in drinking water. The World Health Organisation (WHO), the Nigerian Industrial Standard (NIS), the European Union (EU), and the United State Environmental Protection Agency (USEPA), all developed their drinking water standard. Table 8, shows the general acceptable limits of some metals based on World Health Organisation (WHO) Standard and human health at higher level.

Three different water sources (stream, well and borehole) were analysed for (As, Ba, Ca, Cd, Co, Cr, Cu, Fe, I, K, Mg, Mn, Mo, Na, Ni, P, Pb, S, Sb, Sc, Se, Sr, Ti, V, Z)

Table 5: PH, EC and Temp. Of all the samples measured in the field

Samples ID	Sample Source	Longitude	Latitude	pH	EC (mV)	Temp(°c)
Ada1BH	Borehole	12.176	10.006	7.4	-85.4	36
Ada2BH	Borehole	12.158	10.567	6.6	-35.1	32
Ada3BH	Borehole	12.191	10.592	7.6	-93.8	37
Ada4BH	Borehole	12.19	10.612	7.3	-79.7	26
Ada1W	Well	12.123	10.608	7.5	-89.3	27
Ada2W	Well	12.124	10.61	6.8	-47.8	39
Ada3W	Well	12.24	10.526	7.1	-65.5	37
Ada4W	Well	12.228	10.52	7.4	-86.2	33
Ada5W	Well	12.206	10.608	7.8	-105.3	36
Ada6W	Well	12.207	10.571	7.4	-81.1	33
Ada7W	Well	12.137	10.59	7.4	-82.1	31
Ada8W	Well	12.139	10.589	7.5	-88.3	28
Ada9W	Well	12.14	10.555	7.5	-92.8	33
Ada10W	Well	12.134	10.542	7.2	-73	31
Ada11W	Well	12.153	10.552	7.2	-71	30
Ada12W	Well	12.169	10.619	7	-58.9	26
Ada13W	Well	12.158	10.573	6.3	-17.1	30
Ada14W	Well	12.158	10.569	6.5	-30.9	24
Ada15W	Well	12.201	10.603	6.8	-49.2	34
Ada16W	Well	12.205	10.586	6.7	-42.7	36
Ada17W	Well	12.208	10.583	7	-59.7	28
Ada18W	Well	12.205	10.54	7	-61.3	31
Ada19W	Well	12.202	10.52	6.9	-58.4	25
Ada20W	Well	12.172	10.556	7	-59.1	30
Ada1SW	Surface Water	12.134	10.618	6.9	-54.6	25
Ada2SW	Surface Water	12.175	10.625	7.5	-91	29
Ada3SW	Surface Water	12.191	10.626	7.4	-87.4	30
Ada4SW	Surface Water	12.19	10.609	7.9	-111	38
Ada5SW	Surface Water	12.22	10.621	6.9	-55.4	31
Ada6SW	Surface Water	12.134	10.575	7.5	-89.2	27
Ada7SW	Surface Water	12.172	10.604	7.6	-97.1	19
Ada8SW	Surface Water	12.157	10.601	6.7	-40	29
Ada9SW	Surface Water	12.175	10.6	7.7	-101	40
Ada10SW	Surface Water	12.176	10.54	7.3	-76.7	25
Ada11SW	Surface Water	12.174	10.555	7	-61.7	36
Ada SPW	Spring	12.24	10.525	7.5	-90.9	18
Ada TLK	Tila Lake	12.127	10.542	7.2	-72.6	34

Table 6a: Major and Trace Element Concentration in water

Samples ID	Locality	Coordinates	As (mg/l)	Ba (mg/l)	Ca (mg/l)	Cd (mg/l)	Co (mg/l)	Cr (mg/l)	Cu (mg/l)	Fe (mg/l)	I (mg/l)	K (mg/l)	Mg (mg/l)	Mn (mg/l)
BH1	Waka	10°38'00.3"N, 12°10'50.9"E	0.224	0.025	127.3	0.003	0.002	0.011	0.02	0.005	27.46	<DL	61.83	0.095
BH2	Hena	10°34'02.5"N, 12°09'48.1"E	0.089	0.019	56.27	0.004	0.001	0.001	0.004	<DL	13.59	<DL	32.72	0.039
BH3	Army Barrack	10°35'49.5"N, 12°11'46.4"E	0.039	0.017	73.18	0.002	0.001	0.002	0.006	<DL	<DL	<DL	42.32	0.046
BH4	Biu BCJ	10°36'17.4"N, 12°11'38.7"E	0.07	0.016	48.76	0.002	0.002	0.001	0.005	<DL	12.45	<DL	26.44	0.031
W1	BCJ	10°36'48.4"N, 12°07'40.2"E	0.04	0.012	22.27	<DL	<DL	<DL	<DL	<DL	<DL	<DL	13.45	0.008
W2	Wakama	10°36'57.3"N, 12°07'41.5"E	0.315	0.2	299.7	0.006	0.007	0.022	0.048	0.128	0.24	22.51	171.3	0.313
W3	Yimirshika	10°31'55.8"N, 12°14'38.5"E	0.037	0.014	19.14	<DL	<DL	<DL	<DL	<DL	<DL	<DL	7.346	0.011
W4	Waka	10°31'17.5"N, 12°13'07"E	0.023	0.014	31.9	<DL	<DL	<DL	0.002	<DL	<DL	<DL	14.84	0.021
W5	Biu	10°36'40.5"N, 12°12'40.3"E	0.227	0.181	317.6	0.002	0.012	0.013	<DL	1.516	13.6	7.791	89.89	0.65
W6	biladega	10°34'25.4"N, 12°12'47"E	0.066	0.037	62.83	0.001	0.002	0.001	0.009	<DL	1.778	<DL	25.33	0.051
W7	Tabra Fulani	10°35'35.4"N, 12°07'57.7"E	0.083	0.037	74.85	0.001	0.001	<DL	0.008	<DL	<DL	<DL	37.58	0.046
W8	Malan	10°35'35.4"N, 12°08'22.8"E	0.039	0.013	42.03	<DL	0.001	<DL	0.005	<DL	2.21	<DL	17.92	0.027
W9	Tila	10°33'31.4"N, 12°08'37.8"E	0.059	0.019	46.05	0.001	0.001	<DL	0.007	<DL	<DL	<DL	28.47	0.035
W10	Tila	10°35'51.6"N, 12°08'05.1"E	0.043	0.012	34.36	0.001	<DL	<DL	0.004	<DL	<DL	<DL	18.44	0.016
W11	Tila	10°33'13.3"N, 12°09'18.2"E	0.006	0.031	52.54	0.002	0.001	0.001	0.017	<DL	10.49	<DL	36.52	0.05
W12	Yimirshika	10°37'11.5"N, 12°10'15.6"E	<DL	<DL	<DL	<DL	<DL	<DL	<DL	<DL	<DL	<DL	<DL	<DL
W13	Hena	10°34'37.4"N, 12°09'46.6"E	0.123	0.025	54.27	<DL	0.002	0.001	0.006	0.032	<DL	2.824	24.17	0.059
W14	Hena	10°34'11.5"N, 12°09'47.3"E	0.229	0.085	254.8	<DL	0.007	0.016	0.048	0.141	8.145	10.8	99.17	0.409
W15	BCJ	10°36'20"N, 12°12'04.2"E	<DL	<DL	<DL	<DL	<DL	<DL	<DL	<DL	<DL	<DL	<DL	<DL
W16	Yimirshika	10°35'14.8"N, 12°12'29.8"E	0.136	0.012	26.4	<DL	0.002	<DL	0.003	0.024	<DL	0.646	9.325	0.042
W17	Gwarta	10°35'N, 12°12'49.8"E	0.29	0.12	171	<DL	0.008	0.015	0.033	0.07	10.59	5.808	60.19	0.163
W18	Kunar	10°32'38.1"N, 12°12'27.6"E	0.424	0.06	282.5	<DL	0.004	0.026	0.06	0.247	11.89	11.13	139.9	0.266
W19	Filin Jirgi	10°31'20.1"N, 12°12'27.6"E	<DL	<DL	<DL	<DL	<DL	<DL	<DL	<DL	<DL	<DL	<DL	<DL
W20	Biladega	10°33'36.6"N, 12°10'31.9"E	<DL	<DL	<DL	<DL	<DL	<DL	<DL	<DL	<DL	<DL	<DL	<DL
SW1	Wakama	10°37'04.6"N, 12°08'00.9"E	0.067	0.062	33.88	0.002	0.001	<DL	0.004	<DL	9.7	<DL	29.52	0.021
SW2	Waka	10°37'49.7"N, 12°10'48.4"E	0.389	0.21	340.1	0.007	0.007	0.023	0.053	0.077	87.62	36.16	198.5	0.207
SW3	Waka	10°37'57.8"N, 12°11'44.1"E	0.087	0.033	53.06	0.001	0.001	<DL	0.007	<DL	<DL	<DL	27.12	0.034
SW4	Tila	10°37'23.1"N, 12°13'18.2"E	<DL	<DL	<DL	<DL	<DL	<DL	<DL	<DL	<DL	<DL	<DL	<DL
SW5	Tabra Fulani	10°34'26.8"N, 12°08'03.2"E	0.041	0.033	67.83	0.001	0.001	0.002	0.017	<DL	<DL	<DL	31.8	0.07
SW6	Hena	10°36'20.9"N, 12°10'32.4"E	0.074	0.046	85.57	0.002	0.002	0.006	0.027	0.027	15.95	5.87	52.08	0.078
SW7	Hena	10°36'05.4"N, 12°09'55.6"E	0.116	0.034	67.77	0.001	0.002	0.002	0.009	<DL	7.537	0.012	35.54	0.041
SW8	Hena	10°36'01.4"N, 12°10'48.1"E	0.067	0.095	75.39	0.002	0.001	0.001	0.013	<DL	<DL	1.011	46.54	0.065
SW9	Army Barrack	10°32'39.1"N, 12°10'52.9"E	0.021	0.008	17.57	<DL	<DL	<DL	<DL	<DL	<DL	<DL	7.749	0.006
SW10	Army Barrack	10°33'36.6"N, 12°10'41.8"E	0.032	0.009	16.92	<DL	<DL	<DL	<DL	<DL	<DL	<DL	6.879	0.012
SW11	Kunar	10°36'55.6"N, 12°11'39.2"E	0.036	0.023	32.56	0.001	0.001	<DL	<DL	<DL	<DL	<DL	16.56	0.017
SPW	Yimirshika	10°31'52.2"N, 12°14'41.4"E	0.477	0.092	332.2	0.048	0.084	0.344	0.649	1.678	144	153.7	17.35	2.901
TLK	Tila	10°31'48.9"N, 12°07'59.6"E	<DL	<DL	<DL	<DL	<DL	<DL	<DL	<DL	<DL	<DL	<DL	<DL

KEY : <DL = Below Detection Limit,   BH = Borehole,   W = Well,   SW = Surface Water,   SPW = spring Water,   TLK = Tila Lake

Table 6b: Major and Trace Element Concentration in water

Samples ID	Locality	Coordinates	Mg (mg/l)	Na (mg/l)	Ni (mg/l)	P (mg/l)	Pb (mg/l)	S (mg/l)	Sb (mg/l)	Sc (mg/l)	Se (mg/l)	Sr (mg/l)	Ti (mg/l)	V (mg/l)
BH1	Waka	10°38'00.3"N, 12°10'50.9"E	0.041	22.56	0.009	2.134	0.165	131.7	0.034	0.001	0.136	0.733	<DL	0.197
BH2	Hena	10°34'02.5"N, 12°09'48.1"E	0.009	25.64	0.001	0.447	0.107	0.416	0.012	<DL	0.101	0.447	<DL	<DL
BH3	Army Barrack	10°35'49.5"N, 12°11'46.4"E	0.009	30.99	0.003	0.505	0.091	8.458	0.019	<DL	0.047	0.466	<DL	<DL
BH4	Biu BCJ	10°36'17.4"N, 12°11'38.7"E	0.007	77.63	0.003	0.477	0.067	6.036	0.008	<DL	0.11	0.325	<DL	<DL
W1	BCJ	10°36'48.4"N, 12°07'40.2"E	<DL	67.89	0.002	0.322	<DL	<DL	0.019	<DL	0.026	0.138	<DL	<DL
W2	Wakama	10°36'57.3"N, 12°07'41.5"E	0.065	49.79	0.009	2.387	0.211	437.7	0.023	0.001	0.262	2.477	<DL	0.221
W3	Yimirshika	10°31'55.8"N, 12°14'38.5"E	<DL	7.407	0.002	0.169	<DL	<DL	0.017	<DL	0.015	0.103	<DL	<DL
W4	Waka	10°31'17.5"N, 12°13'07" E	<DL	29.65	0.003	0.402	<DL	<DL	0.018	<DL	0.017	0.182	<DL	<DL
W5	Biu	10°36'40.5"N, 12°12'40.3"E	0.041	19.72	0.033	1.994	0.134	183.8	0.023	0.001	0.245	1.188	<DL	<DL
W6	biladega	10°34'25.4"N, 12°12'47" E	0.004	2.4	0.004	0.545	0.01	7.466	0.017	<DL	0.058	0.358	<DL	<DL
W7	Tabra Fulani	10°35'35.4"N, 12°07'57.7"E	0.007	42.14	0.004	0.979	0.01	2.493	<DL	<DL	0.057	0.533	<DL	<DL
W8	Malan	10°35'35.4"N, 12°08'22.8"E	<DL	23.44	0.004	0.341	<DL	<DL	0.016	<DL	0.044	0.24	<DL	<DL
W9	Tila	10°33'31.4"N, 12°08'37.8"E	0.006	36.67	0.003	0.466	0.001	<DL	0.022	<DL	0.036	0.28	<DL	<DL
W10	Tila	10°35'51.6"N, 12°08'05.1"E	<DL	38.86	0.003	0.496	<DL	<DL	0.019	<DL	0.028	0.182	<DL	<DL
W11	Tila	10°33'13.3"N, 12°09'18.2"E	0.002	17.12	0.003	0.484	0.037	6.785	0.007	<DL	0.055	0.36	<DL	<DL
W12	Yimirshika	10°37'11.5"N, 12°10'15.6"E	<DL	6.858	0.001	<DL	<DL	<DL	0.018	<DL	<DL	<DL	<DL	<DL
W13	Hena	10°34'37.4"N, 12°09'46.6"E	0.013	17.81	0.002	1.005	0.017	<DL	0.015	<DL	0.042	0.444	<DL	<DL
W14	Hena	10°34'11.5"N, 12°09'47.3"E	0.065	23.53	0.009	3.305	0.212	37.33	0.012	0.001	0.271	1.736	<DL	0.322
W15	BCJ	10°36'20"N, 12°12'04.2"E	<DL	<DL	<DL	<DL	<DL	<DL	0.025	<DL	<DL	<DL	<DL	<DL
W16	Yimirshika	10°35'14.8"N, 12°12'29.8"E	0.006	9.561	0.002	0.593	<DL	<DL	0.021	<DL	<DL	0.136	<DL	<DL
W17	Gwarta	10°35'N, 12°12'49.8"E	0.058	16.84	0.011	1.347	0.178	34.44	0.05	0.001	0.243	1.162	<DL	0.763
W18	Kunar	10°32'38.1"N, 12°12'27.6"E	0.076	18.83	0.008	3.591	0.308	58.11	0.035	0.002	0.278	1.864	<DL	0.673
W19	Filin Jirgi	10°31'20.1"N, 12°12'27.6"E	<DL	<DL	<DL	<DL	<DL	<DL	0.043	<DL	<DL	<DL	0.004	<DL
W20	Biladega	10°33'36.6"N, 12°10'31.9"E	<DL	<DL	<DL	<DL	<DL	<DL	0.021	<DL	0.039	0.267	0.056	<DL
SW1	Wakama	10°37'04.6"N, 12°08'00.9"E	0.004	27.8	0.003	0.549	0.033	<DL	0.015	<DL	0.441	2.302	<DL	0.059
SW2	Waka	10°37'49.7"N, 12°10'48.4"E	0.071	53.75	0.023	3.927	0.348	425.8	0.071	0.002	0.056	0.345	<DL	<DL
SW3	Waka	10°37'57.8"N, 12°11'44.1"E	0.005	42.82	0.005	0.559	0.034	<DL	0.021	<DL	<DL	<DL	<DL	<DL
SW4	Tila	10°37'23.1"N, 12°13'18.2"E	<DL	112.7	<DL	<DL	<DL	<DL	0.026	<DL	0.075	0.377	0.001	<DL
SW5	Tabra Fulani	10°34'26.8"N, 12°08'03.2"E	0.004	30.73	0.01	0.748	0.037	16.8	0.025	<DL	0.158	0.724	<DL	<DL
SW6	Hena	10°36'20.9"N, 12°10'32.4"E	0.021	44.12	0.012	1.33	0.081	70.24	0.013	<DL	0.047	0.359	<DL	<DL
SW7	Hena	10°36'05.4"N, 12°09'55.6"E	0.007	42.74	0.002	0.946	0.03	11.2	0.015	<DL	0.025	0.478	<DL	<DL
SW8	Hena	10°36'01.4"N, 12°10'48.1"E	0.007	25.63	0.006	1.006	0.02	4.378	0.02	<DL	0.009	0.107	<DL	<DL
SW9	Army Barrack	10°32'39.1"N, 12°10'52.9"E	<DL	17.17	0.003	0.257	<DL	<DL	0.015	<DL	0.012	0.094	<DL	<DL
SW10	Army Barrack	10°33'36.6"N, 12°10'41.8"E	<DL	48.12	0.003	0.173	<DL	<DL	0.019	<DL	0.054	0.212	<DL	<DL
SW11	Kunar	10°36'55.6"N, 12°11'39.2"E	<DL	33.89	0.001	0.225	<DL	<DL	0.023	<DL	<DL	<DL	<DL	<DL
SPW	Yimirshika	10°31'52.2"N, 12°14'41.4"E	1.039	21.27	0.003	44.6	0.374	784.7	<DL	0.024	0.432	22.96	<DL	15.6
TLK	Tila	10°31'48.9"N, 12°07'59.6"E	<DL	78.73	<DL	<DL	<DL	<DL	0.041	<DL	<DL	<DL	0.001	<DL

KEY : <DL = Below Detection Limit, BH = Borehole, W = Well, SW = Surface Water, SPW = spring Water, TLK = Tila Lake

48

Table 7: Major and Trace Element (Soil)

Samples ID	Locations	MgO (%)	K2O (%)	CaO (%)	TiO2 (%)	Fe2O3 (%)	V (ppm)	Cr (ppm)	Mn (ppm)	Co (ppm)	Ni (ppm)	Cu (ppm)	Zn (ppm)	As (ppm)	Se (ppm)	Mo (ppm)	Cd (ppm)	Sb (ppm)	Pb (ppm)	Be (ppm)
AD1	10°34'0.63"N, 11°51'29.0"E	1.50	0.65	2.09	0.01	1.12	<DL	34.09	406.9	13.48	43.4	13.49	467.3	<DL	<DL	<DL	1.604	<DL	3.479	<DL
AD2	10°31'48.2"N, 12°14'39.0"E	1.76	0.75	2.31	40.01	49.81	112.2	230.3	791.3	45.87	89.03	51.54	495.6	<DL	<DL	<DL	<DL	<DL	16.83	<DL
AD3	10°31'16.8"N, 12°13'06.8"E	2.34	0.80	0.40	68.67	85.0	82.64	440.9	2074	110.8	227.8	92.21	<DL	<DL	<DL	<DL	<DL	<DL	32.56	<DL
AD4	10°37'56.9"N, 12°11'41.9"E	8.18	0.82	4.16	15.31	28.41	<DL	246.7	1153	59.71	296.7	60.65	252.3	<DL	<DL	<DL	<DL	<DL	6.391	<DL
AD5	10°36'51.1"N, 12°12'50.9"E	3.94	0.72	3.60	3.60	19.33	<DL	151.8	1716	60.29	168.7	37.52	261.7	<DL	<DL	<DL	<DL	<DL	7.9	<DL
AD6	10°34'36.4"N, 12°12'48.1"E	4.11	0.53	4.77	27.07	29.87	<DL	151.5	1507	58.13	221.2	47.15	295.3	<DL	<DL	<DL	<DL	<DL	3.021	<DL
AD7	10°35'34.7"N, 12°07'57.9"E	4.40	0.75	4.13	12.98	34.71	<DL	183.7	3352	111.3	231	45.06	245.6	<DL	<DL	<DL	<DL	<DL	13.58	<DL
AD8	10°34'26.8"N, 12°08'03.2"E	4.54	0.70	3.69	7.35	14.26	<DL	154.3	914.5	41.31	177	42.59	495.6	<DL	<DL	<DL	<DL	<DL	5.161	<DL
AD9	10°32'50.3"N, 12°08'01.9"E	10.81	2.32	11.61	4.37	16.29	<DL	166.3	1857	52.97	170.6	49.62	349.4	<DL	<DL	<DL	<DL	<DL	<DL	<DL
AD10	10°32'50.3"N, 12°08'55.3"E	1.52	0.85	1.49	16.15	30.63	23.64	153.3	862.9	41.68	104	34.09	276.1	<DL	<DL	<DL	<DL	<DL	9.556	<DL
AD11	10°36'08.1"N, 12°09'55.6"E	2.43	0.85	2.62	1.63	23.1	<DL	92.27	1771	46.66	98.06	28.01	268.3	<DL	<DL	<DL	<DL	<DL	42.95	<DL
AD12	10°32'45.2"N, 12°12'30.2"E	6.26	3.09	4.29	6.23	41.16	<DL	163.5	25.28	83.9	250.1	59.97	324.6	<DL	<DL	<DL	<DL	<DL	<DL	<DL
AD13	10°32'29.8"N, 12°11'13.7"E	1.72	0.63	1.92	0.80	13.21	<DL	114.9	8.11	46.61	95.93	29.1	346.6	<DL	<DL	<DL	0.365	<DL	6.803	<DL

<DL = Below Detection Limit

Table 8: World Health Organization Standard for Drinking Water (2008) and Human Health Impact at Higher Level

Elements	WHO (mg/l)	Human Health Impacts At Higher Level.
Arsenic	0.01	Toxic, liver and kidney damage.
Cadmium	0.003	Toxic, causes high blood pressure, liver and kidney damage, anaemia and destroys testicular tissue.
Lead	0.01	Severe and permanent brain damage, can cause slight increase in blood pressure in some adults.
Seleniu	0.01	Toxic, causes blood staggers in animals, development of cancers malformation of nails and hair, depression and nervousness in humans.
Molybdenum	0.07	-
Antimony	0.005	Increase in blood cholesterol; decrease in blood sugar.
Calcium	200	Scale formation.
Magnesium	500	Scale and test.
Sodium	200	Taste
Phosphorous	-	-
Barium	0.3	Increase in blood pressure
Cobolt	-	-
Chromium	0.1	Allergic dermatitis.
Copper	0.05	Taste and liver damage
Iron	0.1	Stain and taste
Iodine	-	Extremely high dose cause abnormal thyroid formation.
Manganese	0.5	Stains, taste and deposit in water lines cause problems of blockage.
Nickel	0.02	Cancer
Strontium	-	-
Zinc	0.1	Bitter taste, gastrointestinal problems and opalescence alkaline water.

Table 9: Average Major and Trace Element Composition of Basalt

Elements	Unit Symbol	1	2
MgO	%	11	10.6
K2O	%	10.2	2.07
CaO	%	22	9.95
TiO2	%	10.3	2.23
Fe_2O_3	%	30	10.71
M_nO	%	0.3	1,500
V	ppm	73	250
Cr	ppm	176	170
Co	ppm	59	60
Ni	ppm	176	100
Cu	ppm	45	87
Zn	ppm	339	105
Pb	ppm	14	6
As	ppm	<DL	2
Cd	ppm	1	-
Se	ppm	-	-
Mo	ppm	-	-
Sb	ppm	-	-

1. Average Major and Trace Element Composition of Biu Plateau Volcanic Soils (This study)
2. Average chemical Composition of Parent Rock (ACCPR) of Biu Plateau Basalt (Saidu, 2004)

Table 10: Average concentration of Major and Trace Element in surface, well, borehole and soil

Elements	Surface Water	Ground Water	Soil	
	(mg/L)			%
Ti	-	0.0015	TiO	20
Fe	0.9	0.0155	Fe$_2$O$_3$	46
Ca	38	94	CaO	126
K	39	4.5	K$_2$O	173
Mg	29	45.5	MgO	123
				ppm
Na	45	32	-	
P	5	0.95	-	
S	22	66.5	-	
As	0.49	0.12	-	
Cd	0.01	0.003	0.02	
Co	0.01	0.003	59	
Cr	0.01	0.15	176	
Cu	0.1	0.11	45	
I	53	13	-	
Mn	0.3	0.10	15	
Mo	0.14	0.025	-	
Ni	0.01	0.005	17	
Pb	0.5	0.1	14	
Sb	0.03	0.01	-	
Sc	0.01	0.001	-	
Se	0.48	0.11	-	
Sr	3	0.6	-	
Ba	0.13	0.4	-	
V	8	0.25	73	
Zn	-	6	34	

WHO

WHO

WHO

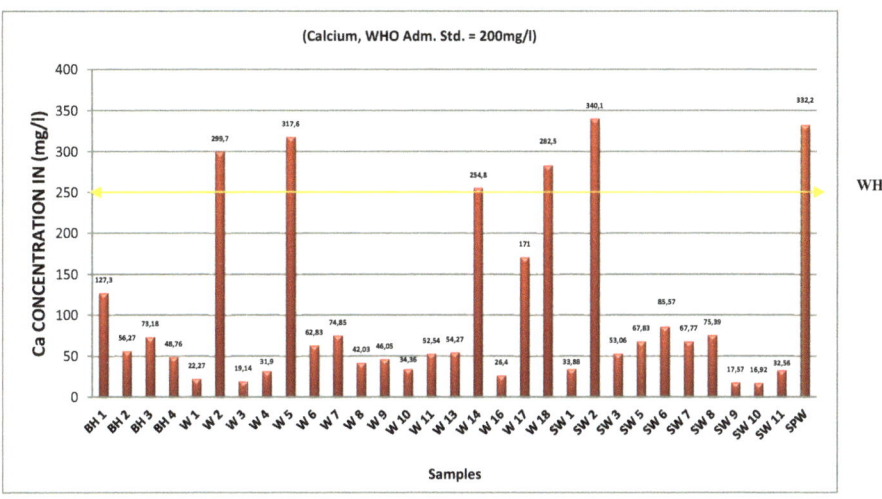

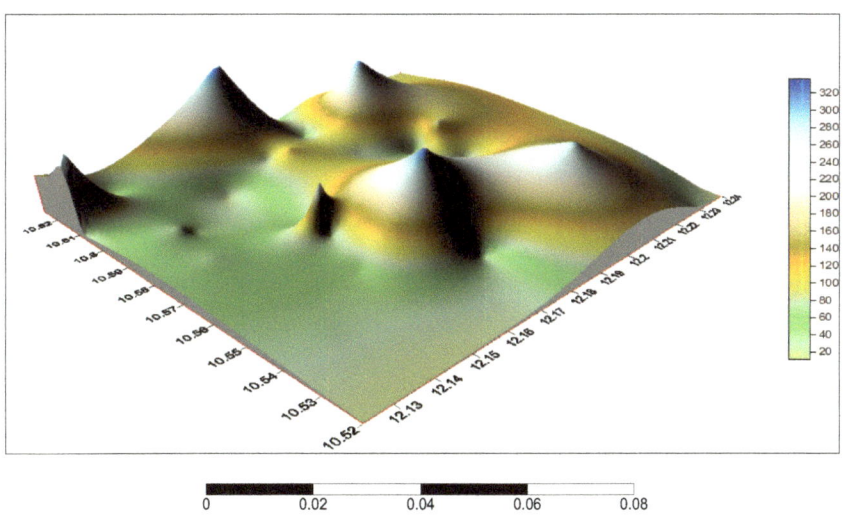

Figure 11: Ca Concentrations in Boreholes (BH) Wells (W) and Surface Water (SW) displayed in Bar Chat and 3D Surface Map

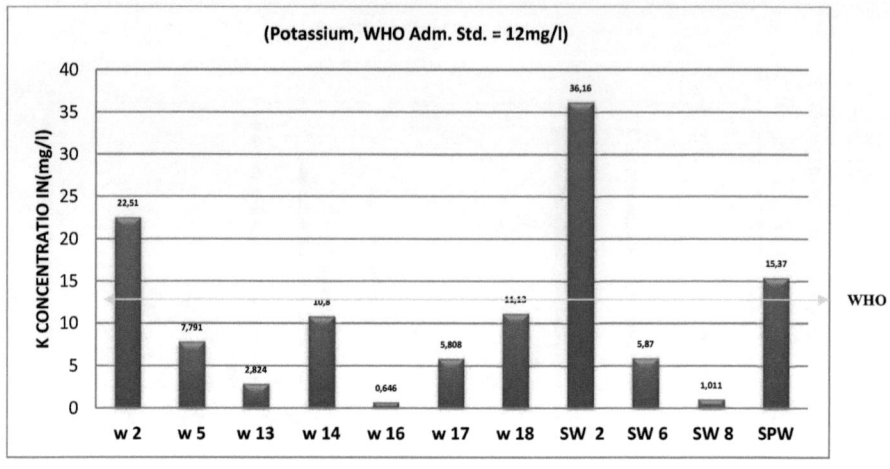

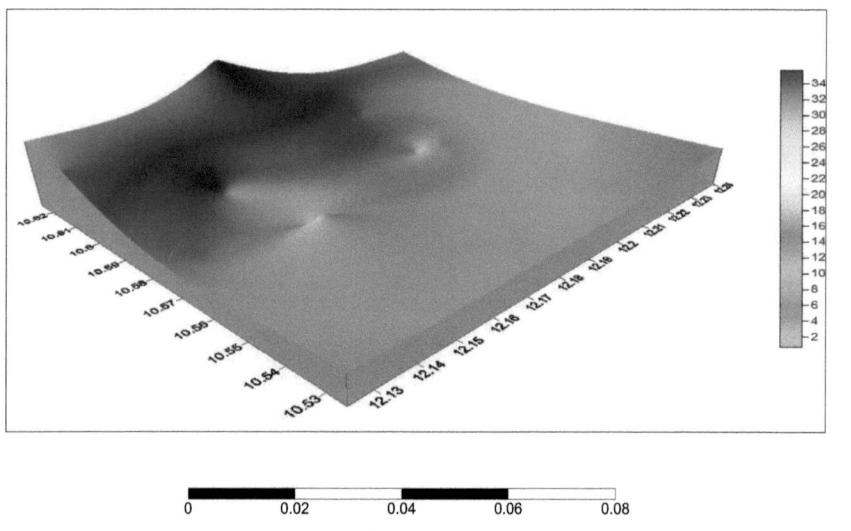

Figure 12: K concentrations in wells and surface water in Bar Chat and 3D Surface Map

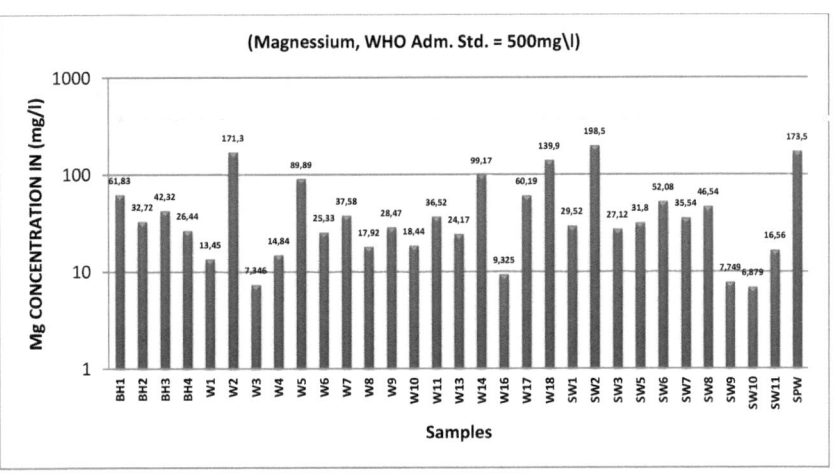
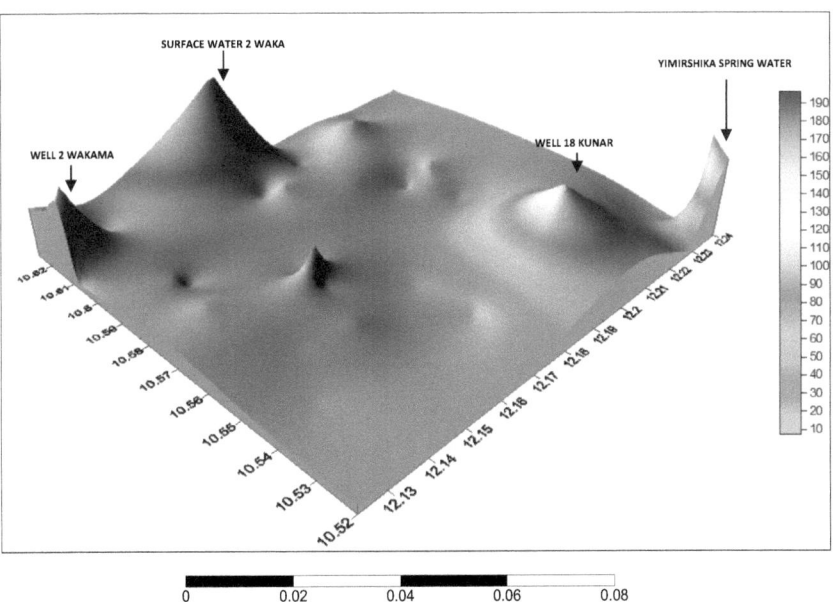

Figure 13: Mg concentrations in wells and surface water displayed in Bar Chat and 3D Surface Map

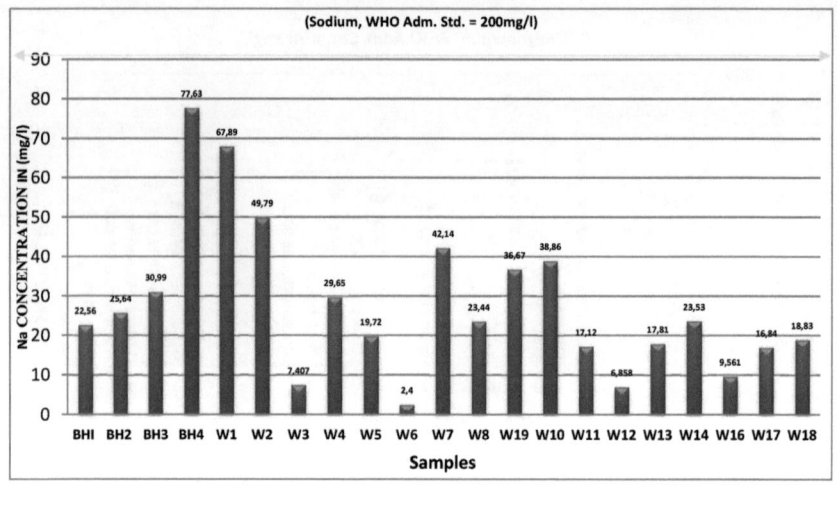

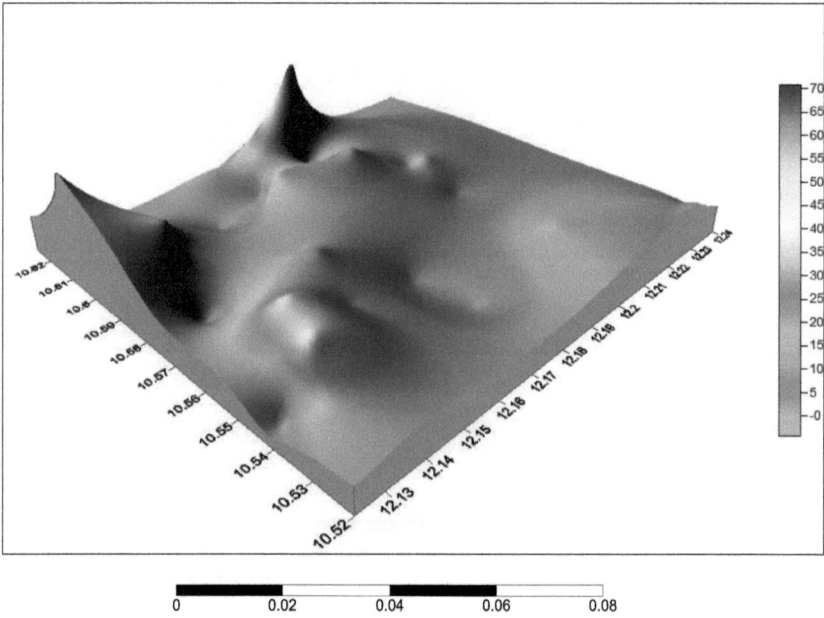

Figure 14: Na concentrations in wells and surface water displayed in Bar Chat and 3D Surface Map

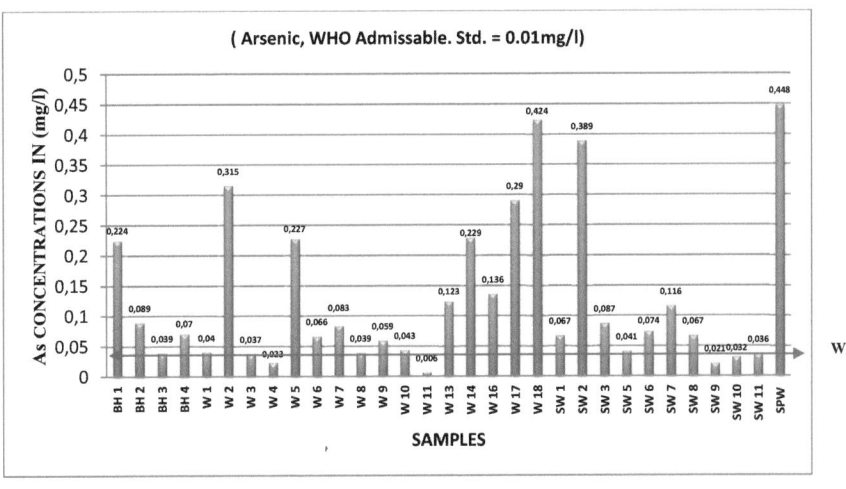

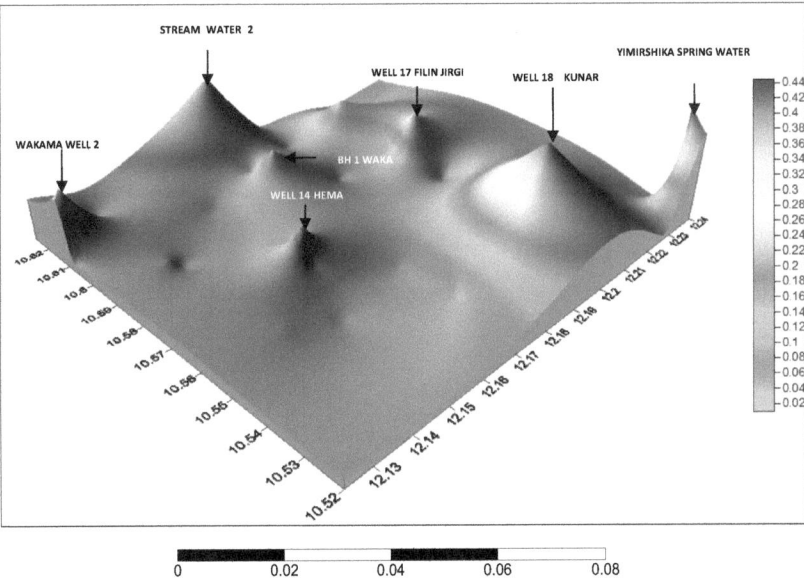

Figure 15: As concentrations in boreholes, wells and surface water displayed in Bar Chat and 3D Surface Map

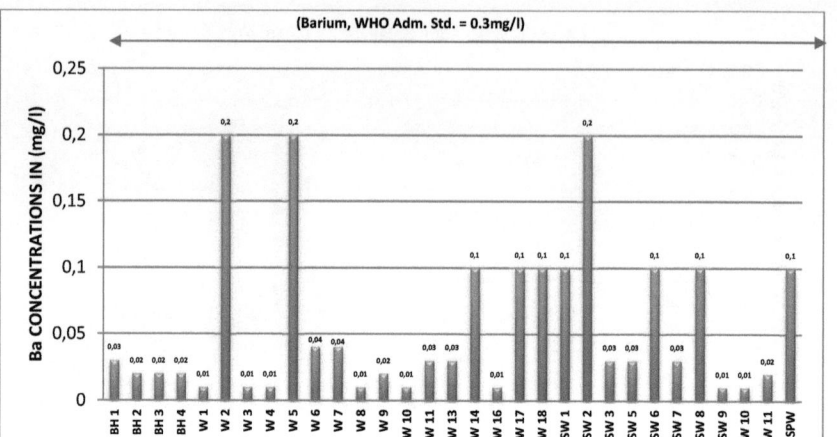
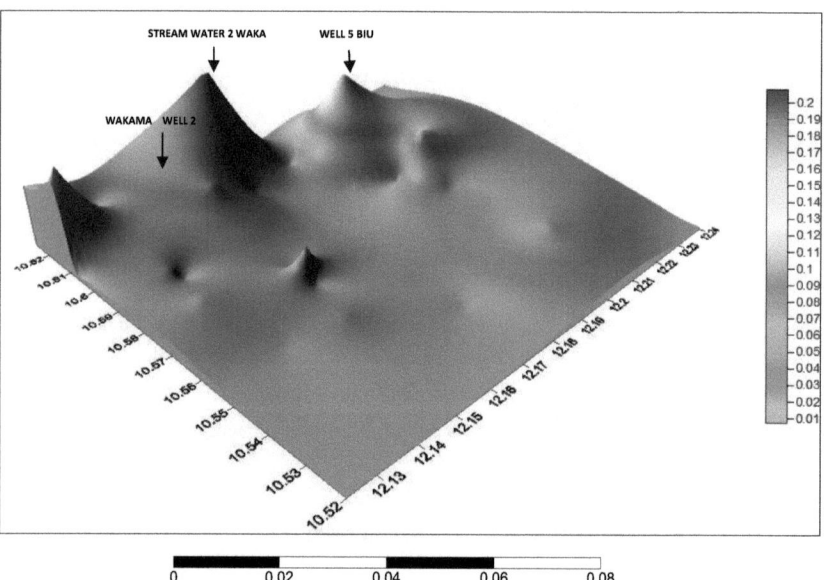

Figure 16: Ba concentrations in boreholes, wells and surface water displayed in Bar Chat and 3D Surface Map

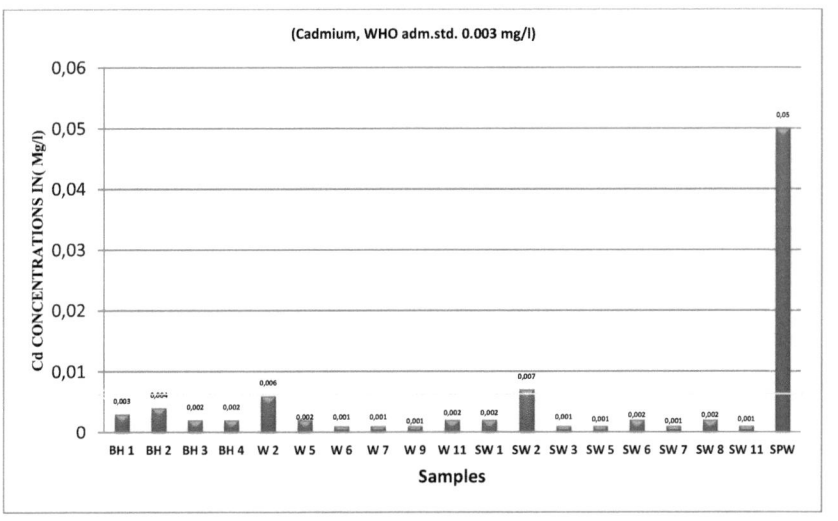
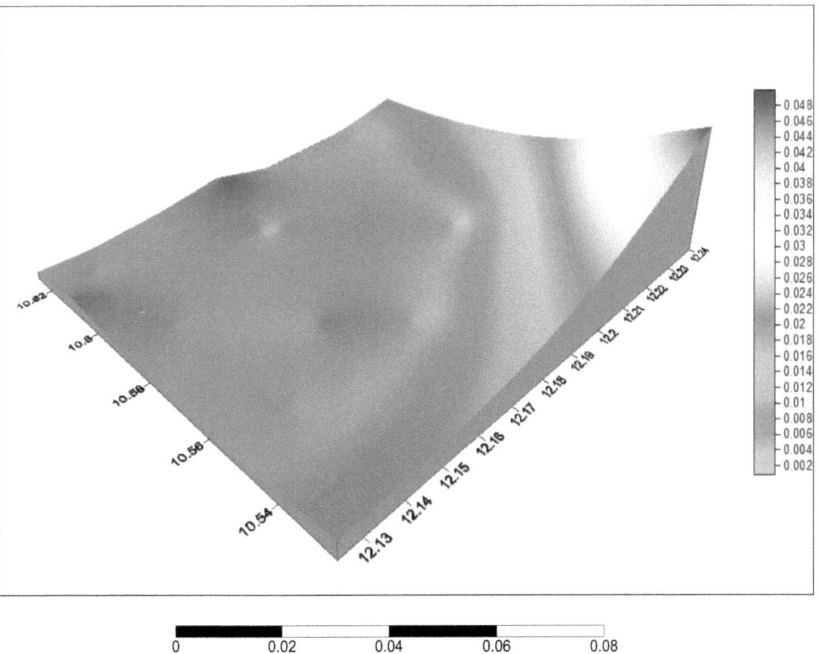

Figure 17: Cd concentrations in boreholes, wells and surface water displayed in Bar Chat and 3D Surface Map

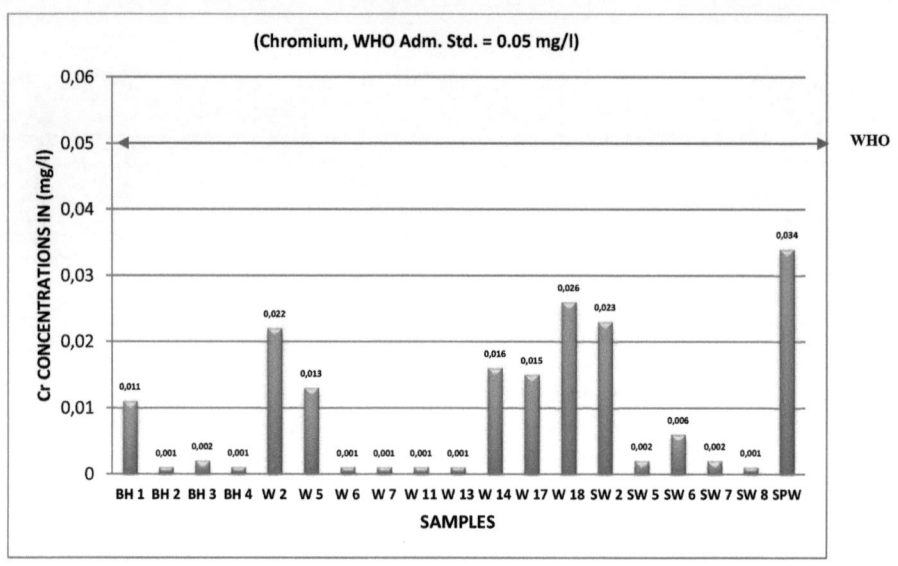

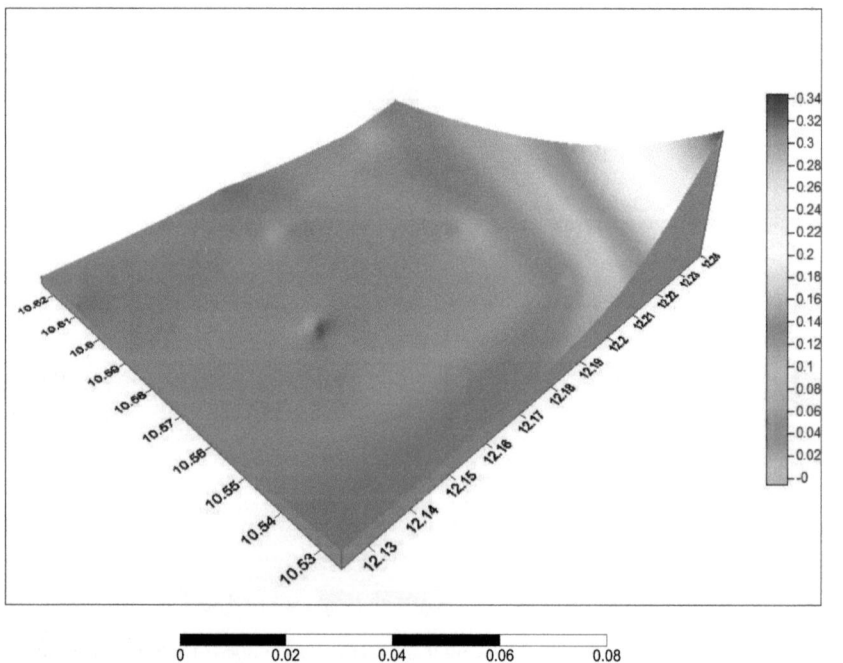

Figure 18: Cr concentrations in boreholes, wells and surface water displayed in Bar Chat and 3D Surface Map

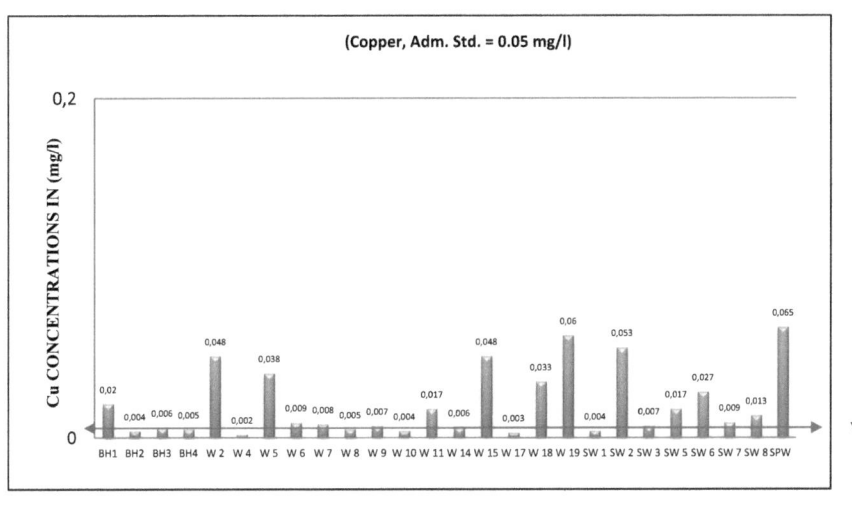

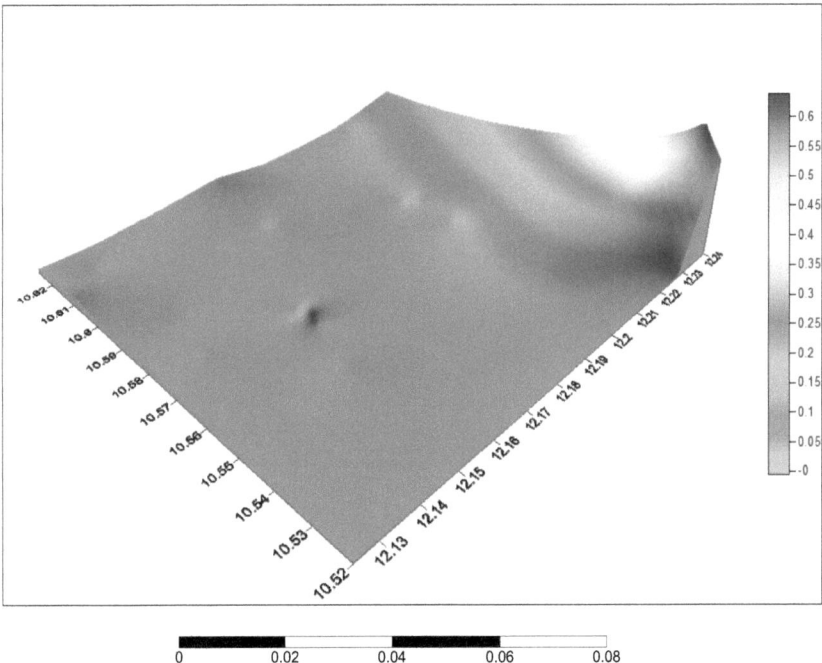

Figure 19: Cu concentrations in boreholes, wells and surface water displayed in Bar Chat and 3D Surface Map

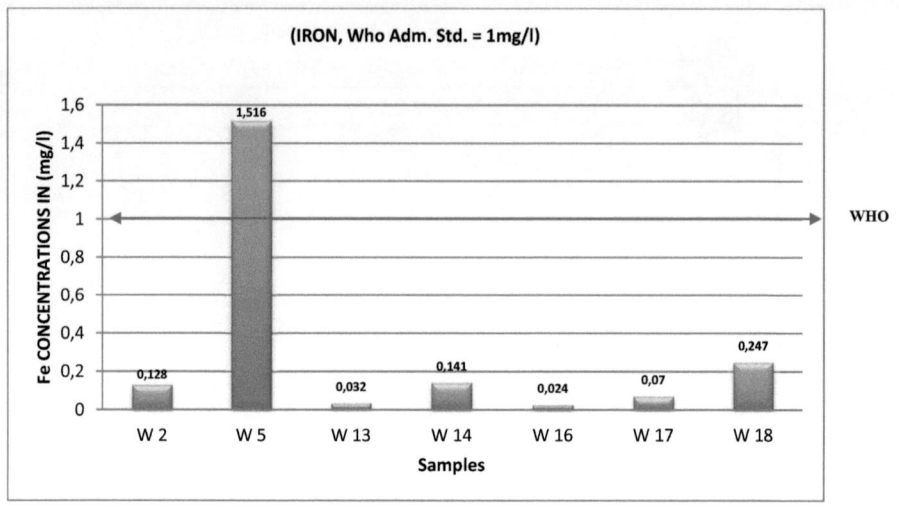

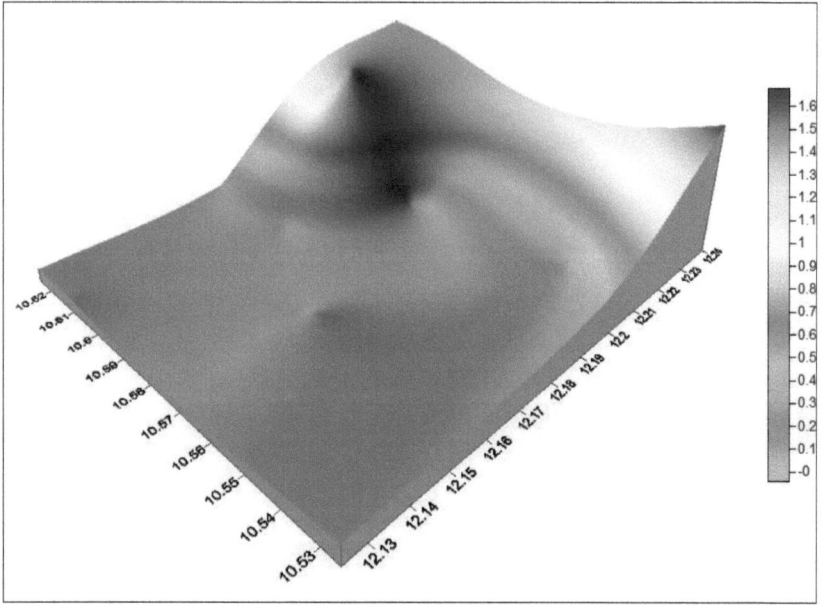

Figure 20: Fe concentrations in boreholes, wells and surface water displayed in Bar Chat and 3D Surface Map

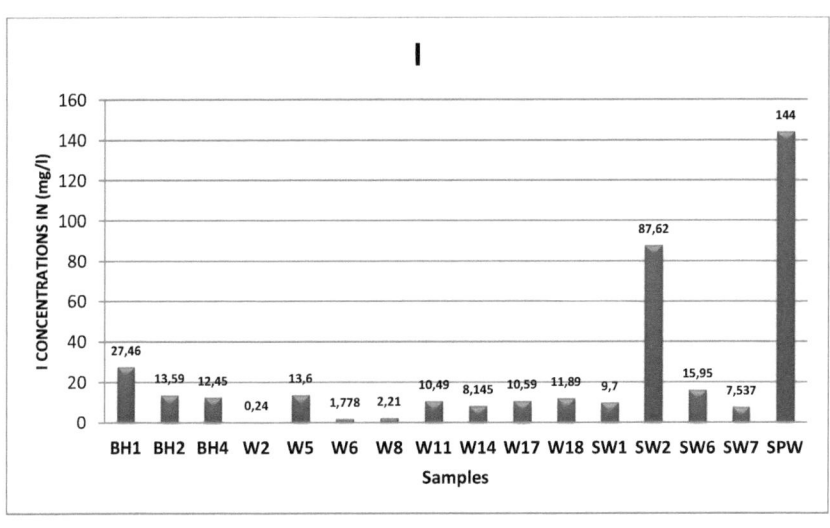
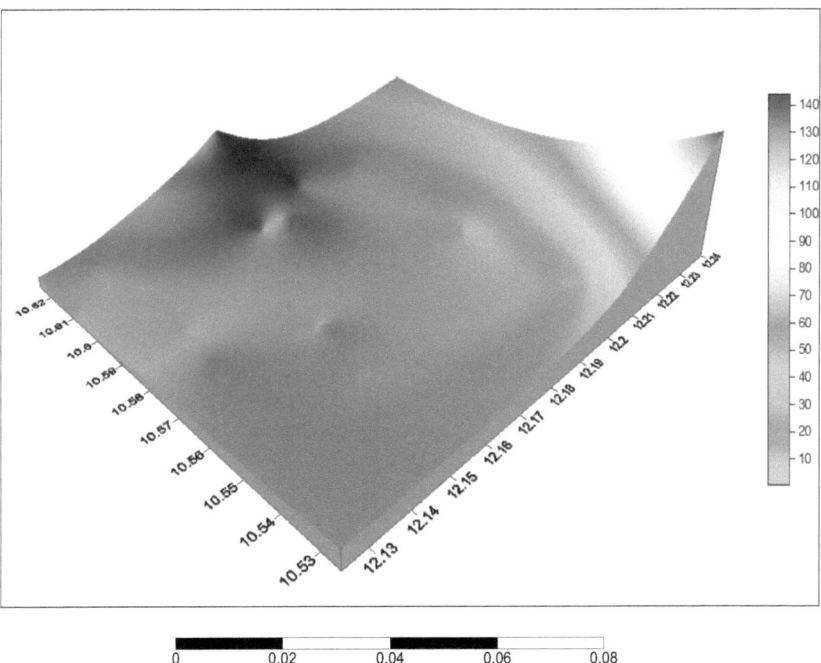

Figure 21: I concentrations in boreholes, wells and surface water displayed in Bar Chat and 3D Surface Map

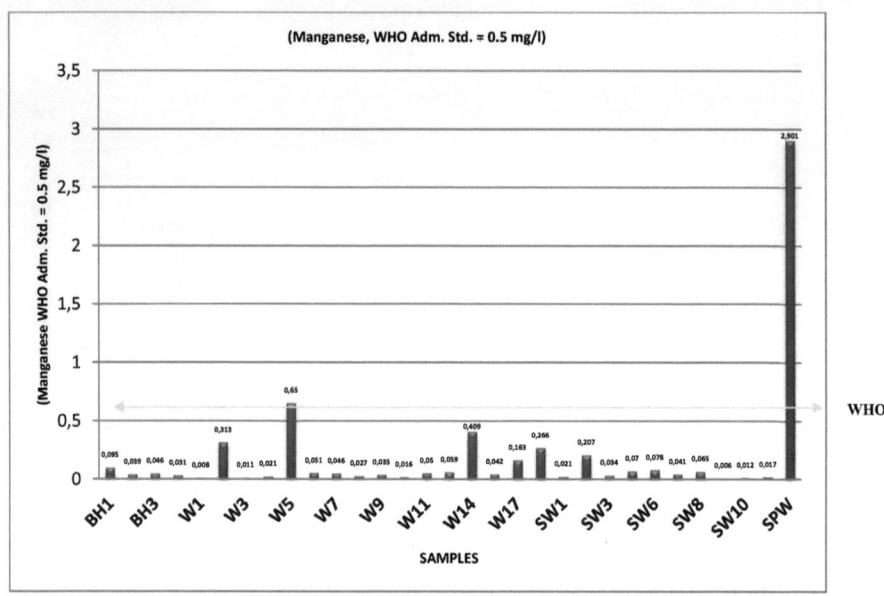

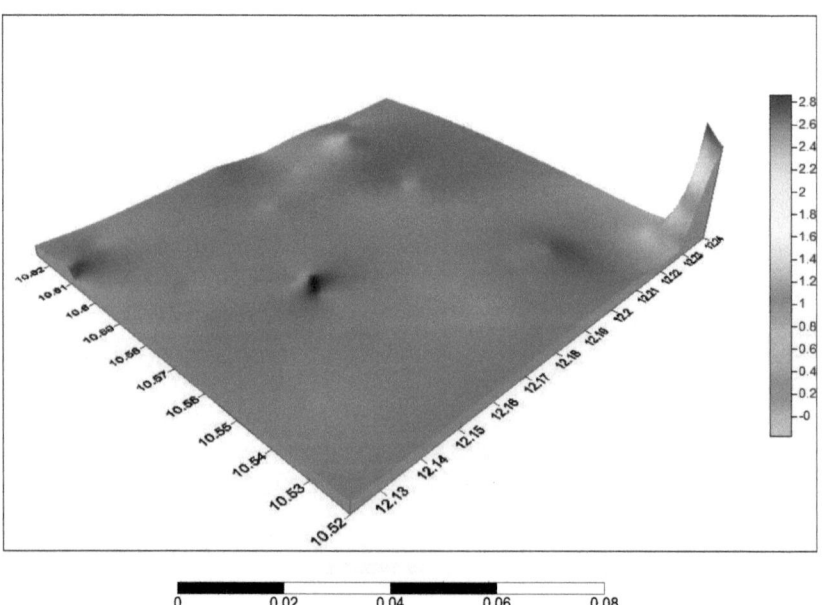

Figure 22: Mn concentrations in boreholes, wells and surface water displayed in Bar Chat and 3D Surface Map

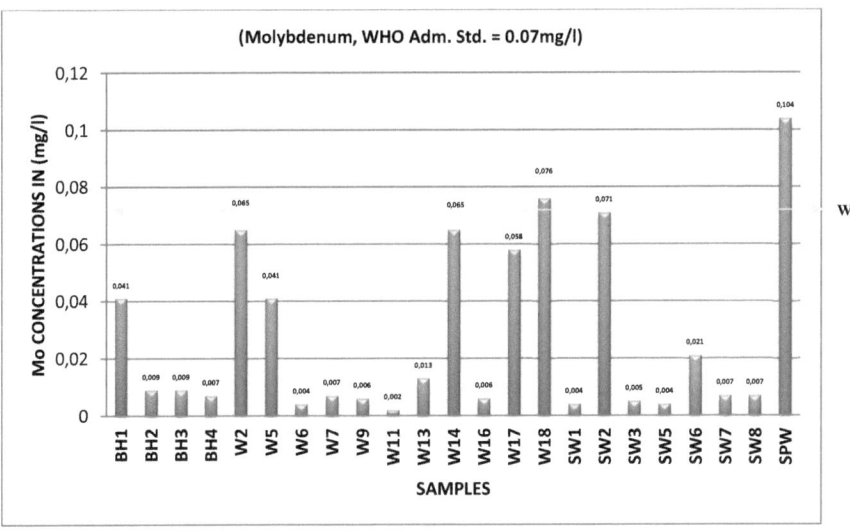

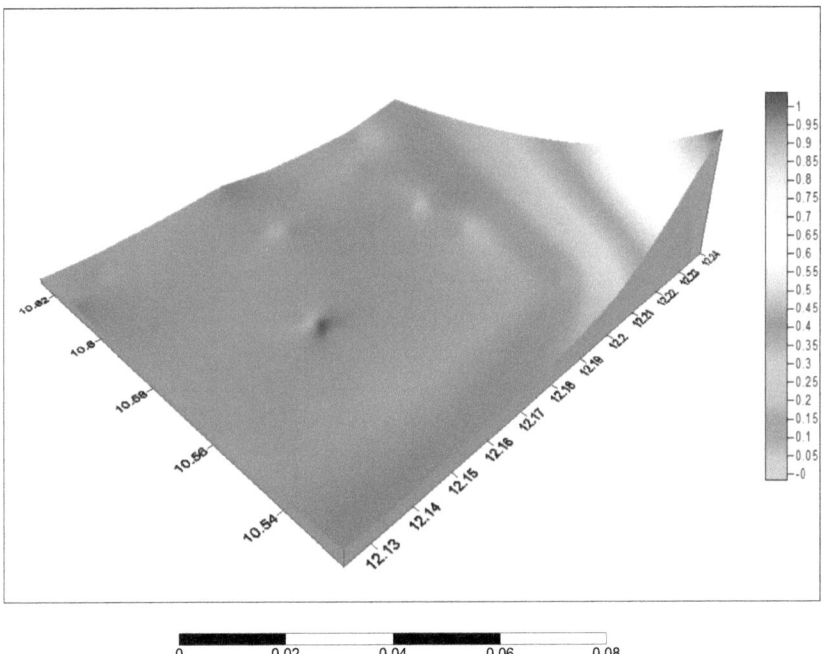

Figure 23: Mo concentrations in boreholes, wells and surface water displayed in Bar Chat and 3D Surface Map

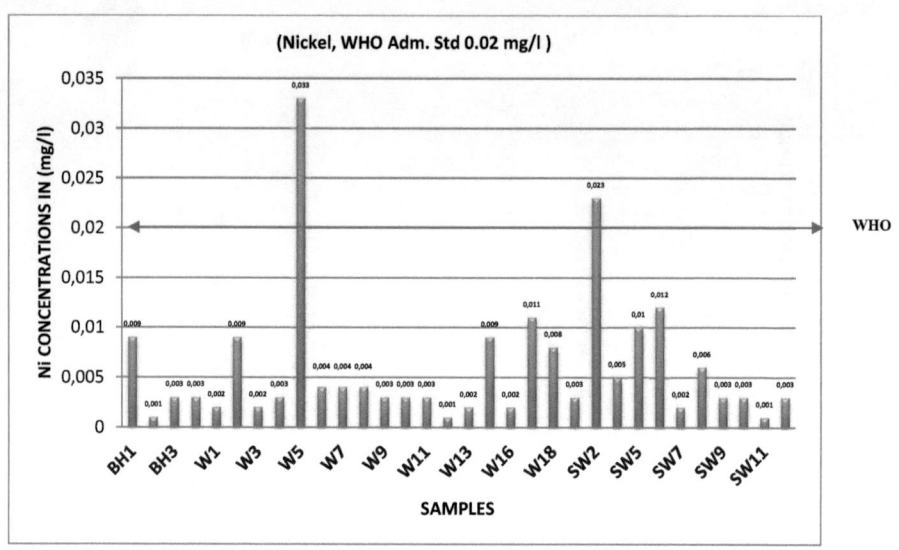

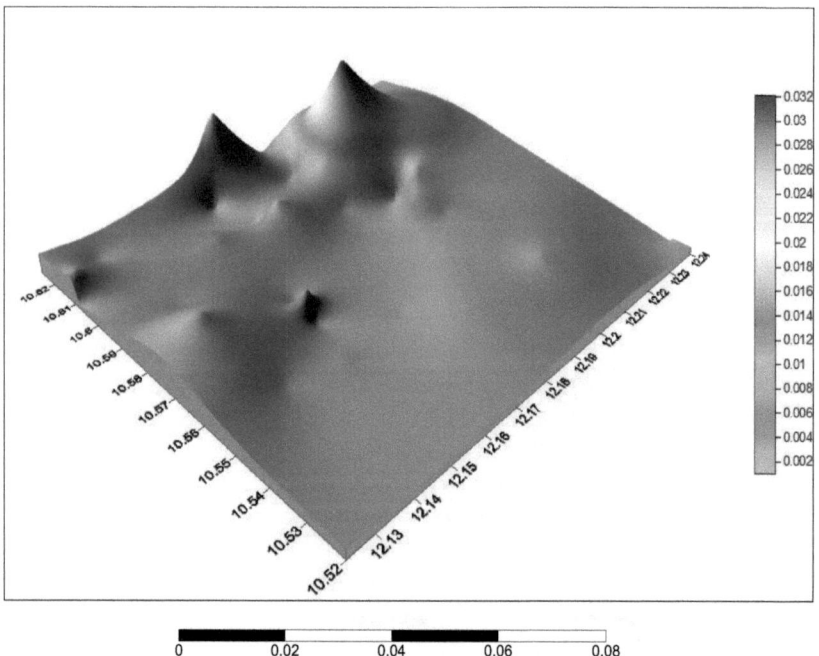

Figure 24: Ni concentrations in boreholes, wells and surface water displayed in Bar Chat and 3D Surface Map

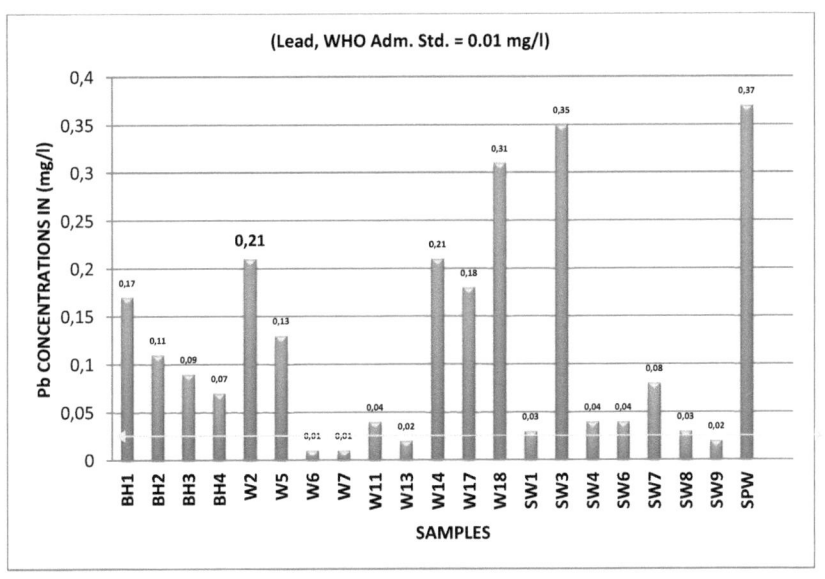
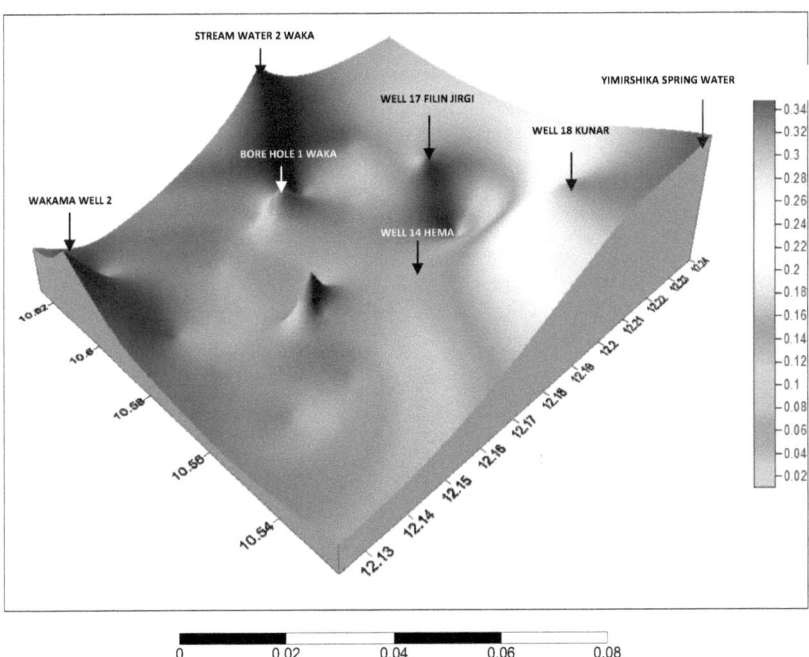

Figure 25: Pb concentrations in boreholes, wells and surface water displayed in Bar Chat and 3D Surface Map

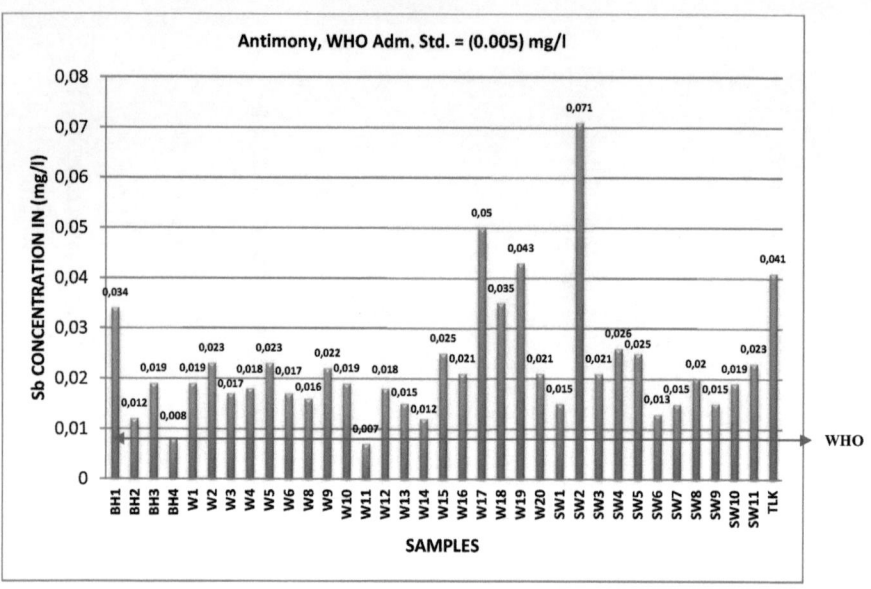
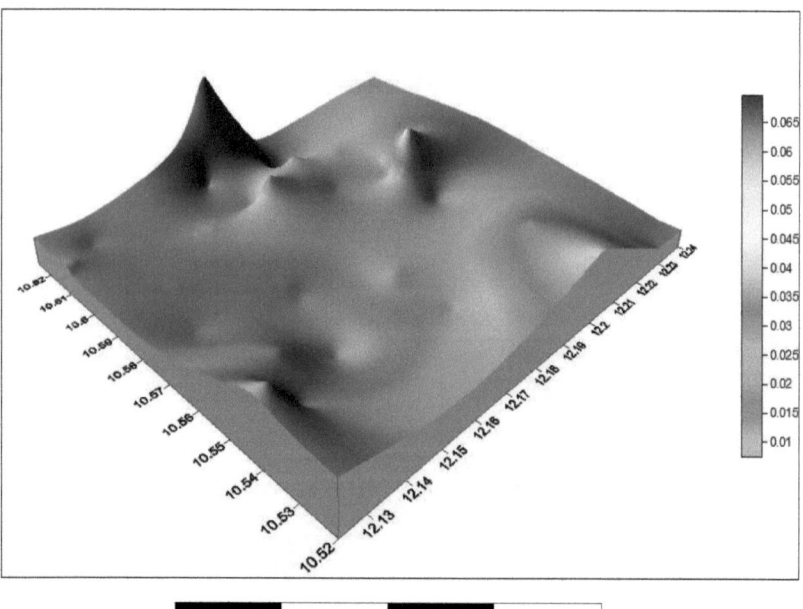

Figure 26: Sb concentrations in boreholes, wells and surface water displayed in Bar Chat and 3D Surface Map

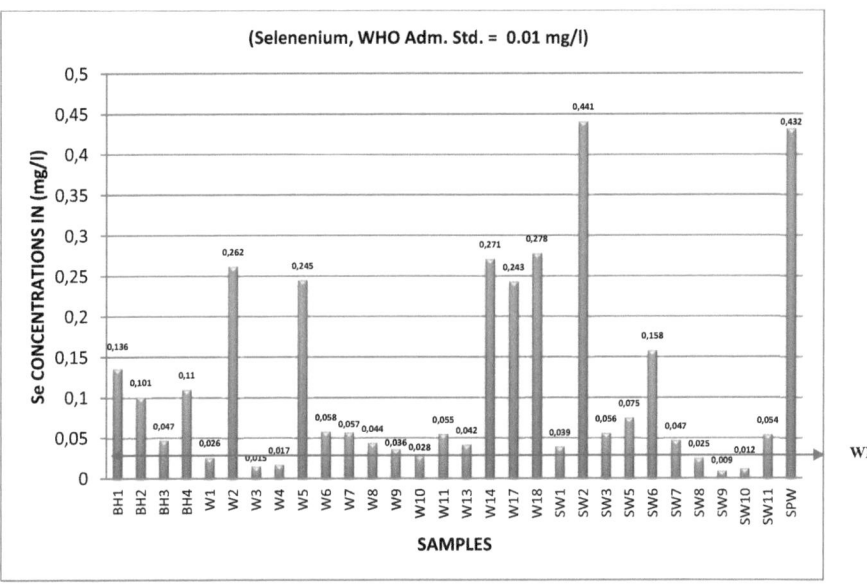
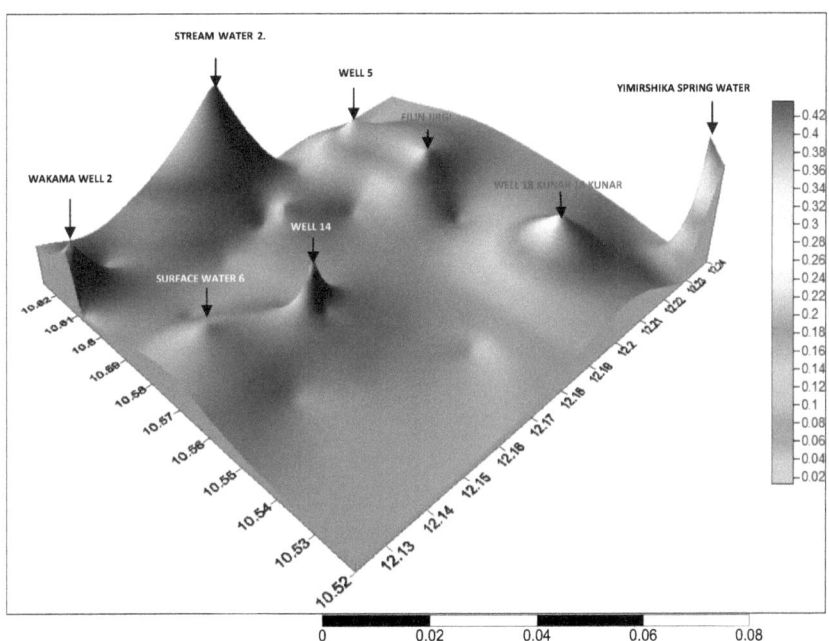

Figure 27: Se concentrations in borehole, wells and surface water displayed in Bar Chat and 3D Surface Map

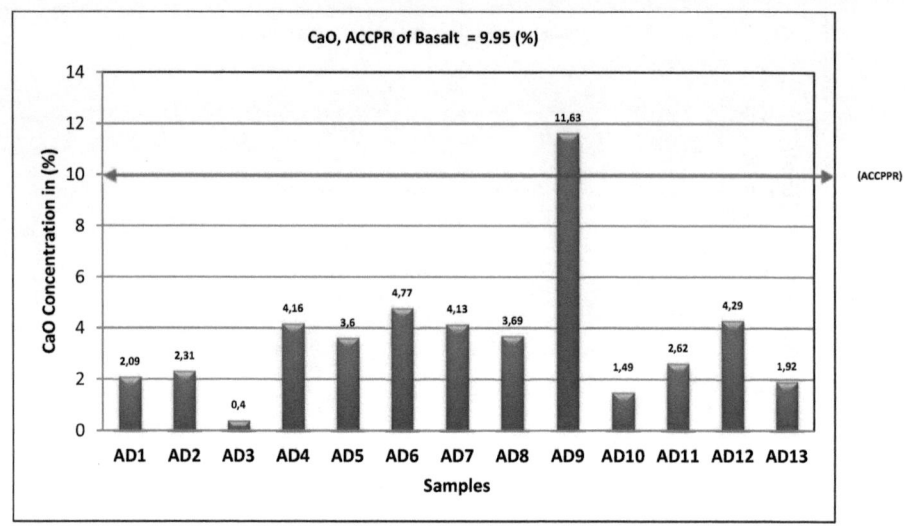

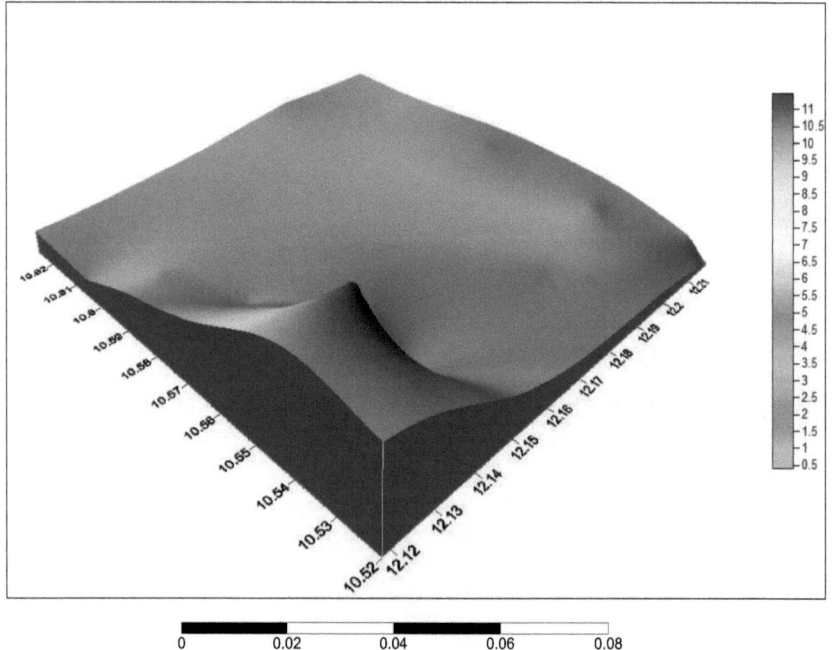

Figure 28: CaO Concentrations in Biu Volcanic Soils in Bar Chat and 3D Surface Map

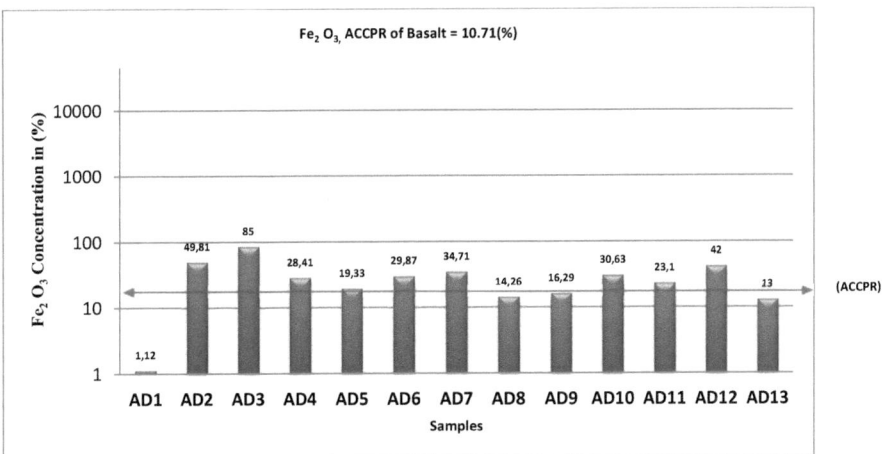

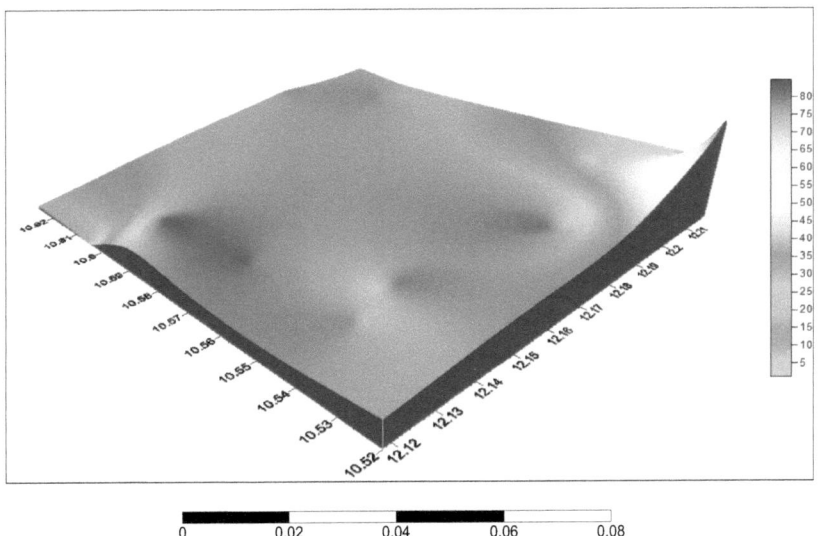

Figure 29: Fe_2O_3 Concentrations in Biu Volcanic Soils in Bar Chat and 3D Surface Map

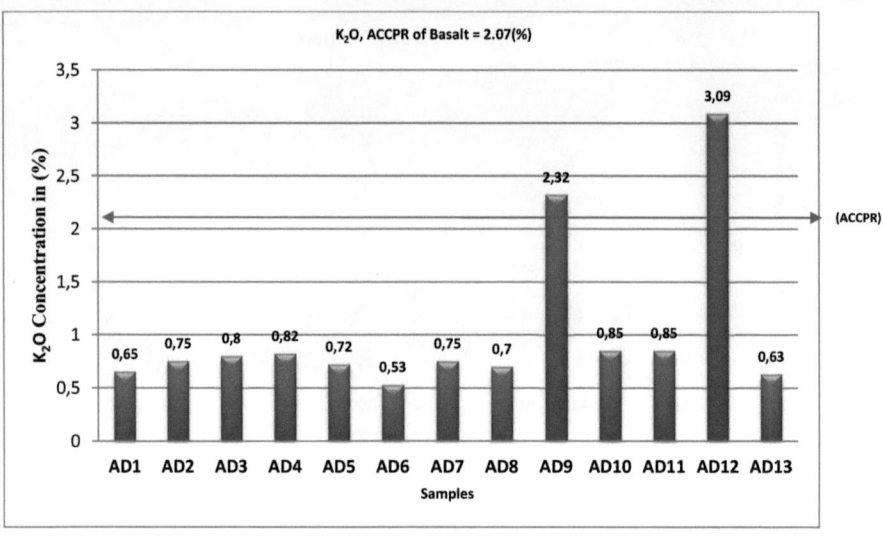

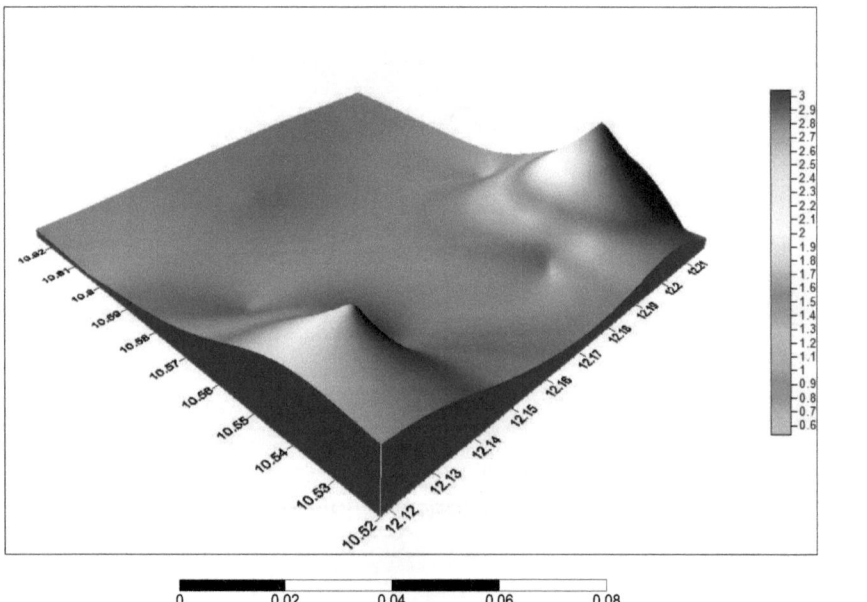

Figure 30: K$_2$O Concentrations in Biu Volcanic Soils in Bar Chat and 3D Surface Map

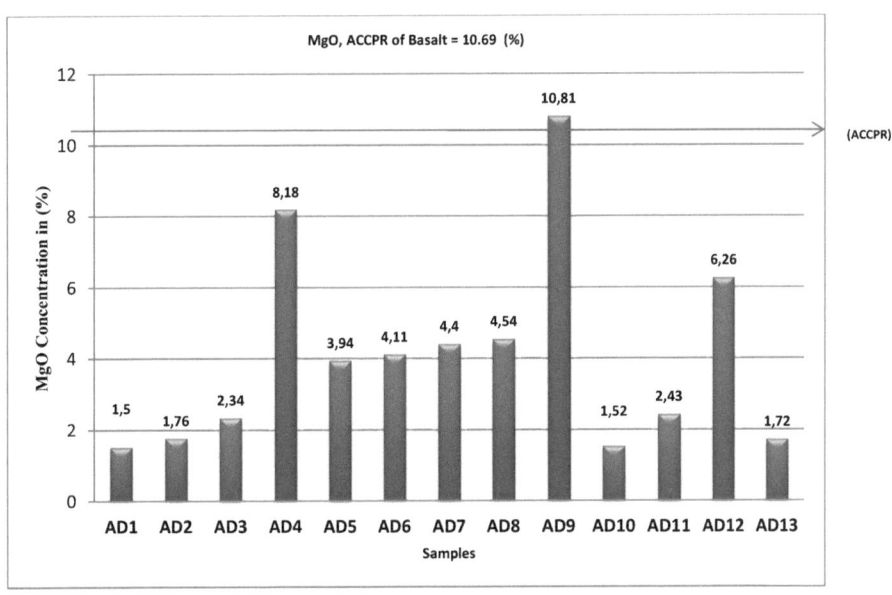
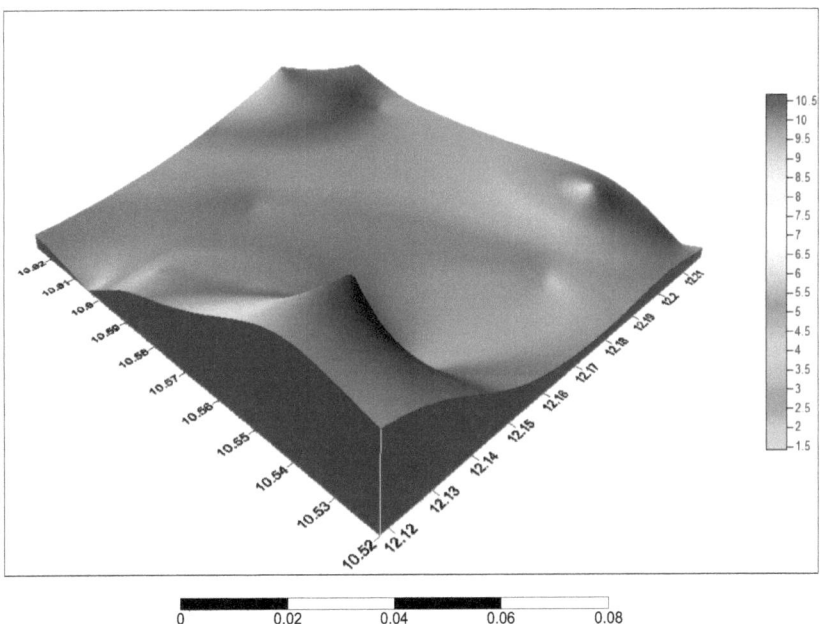

Figure 31: MgO Concentrations in Biu Volcanic Soils in Bar Chat and 3D Surface Map

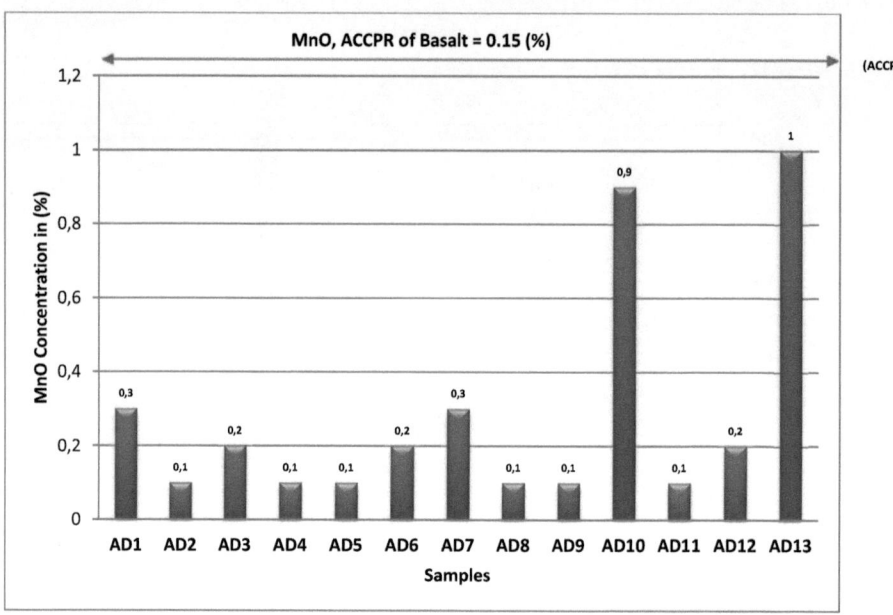
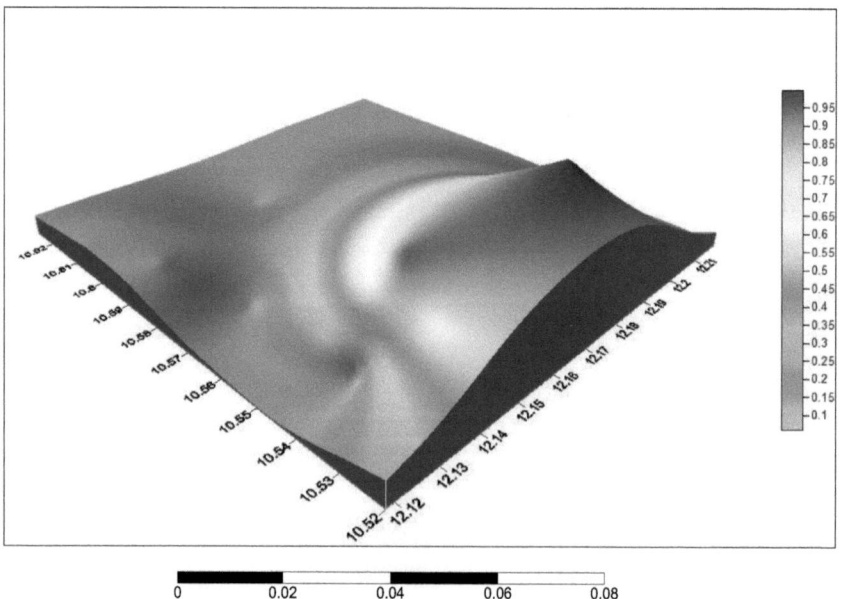

Figure 32: MnO Concentrations in Biu Volcanic Soils in Bar Chat and 3D Surface Map

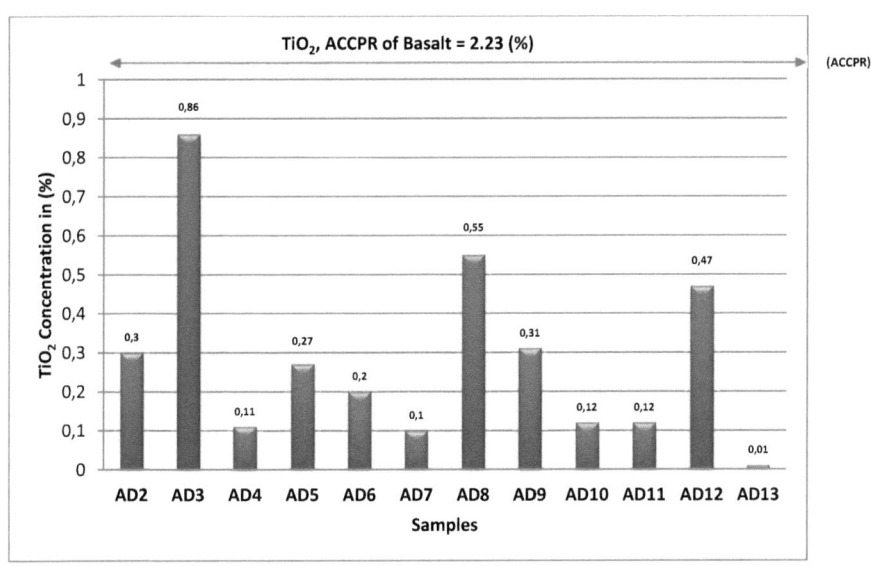

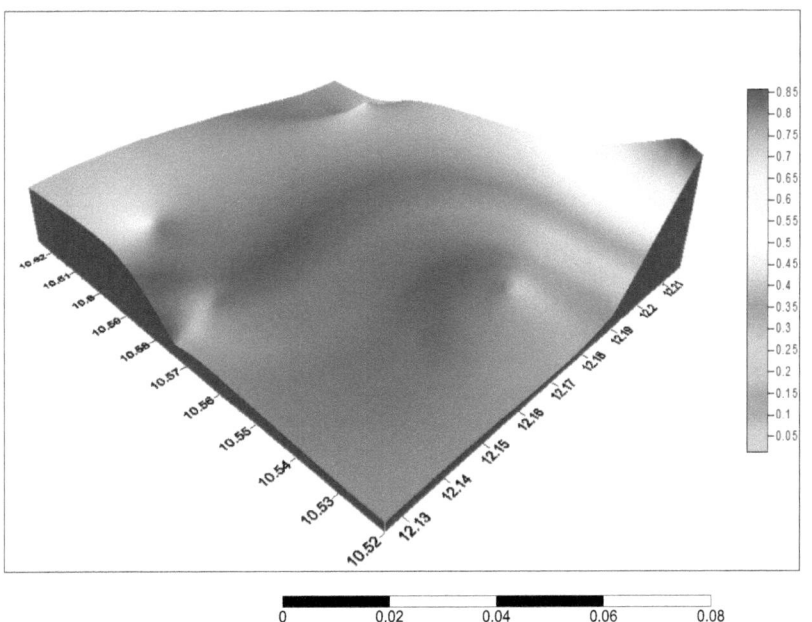

Figure 33: TiO Concentrations in Biu Volcanic Soils in Bar Chat and 3D Surface Map

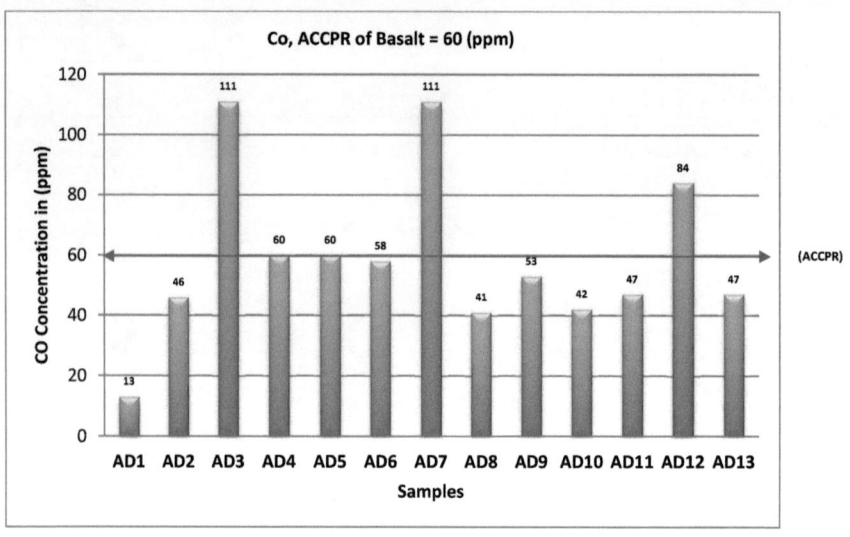
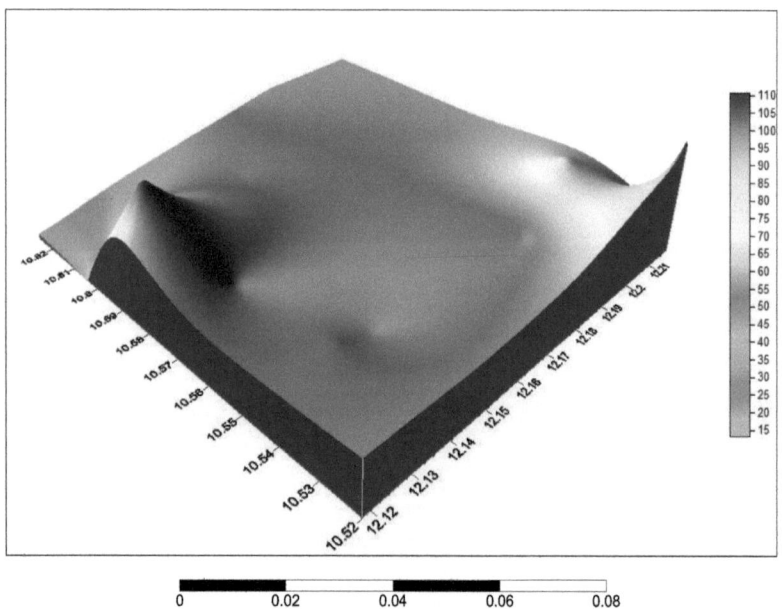

Figure 34: Co Concentrations in Biu Volcanic Soils in Bar Chat and 3D Surface Map

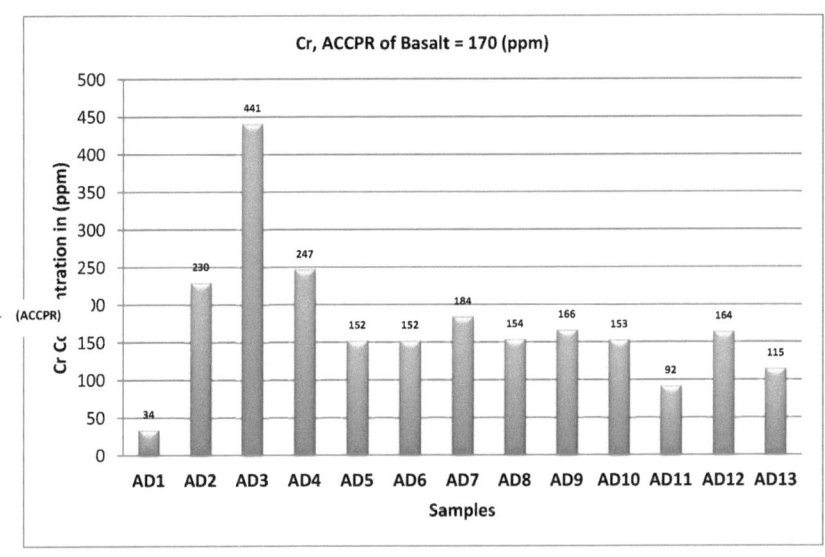

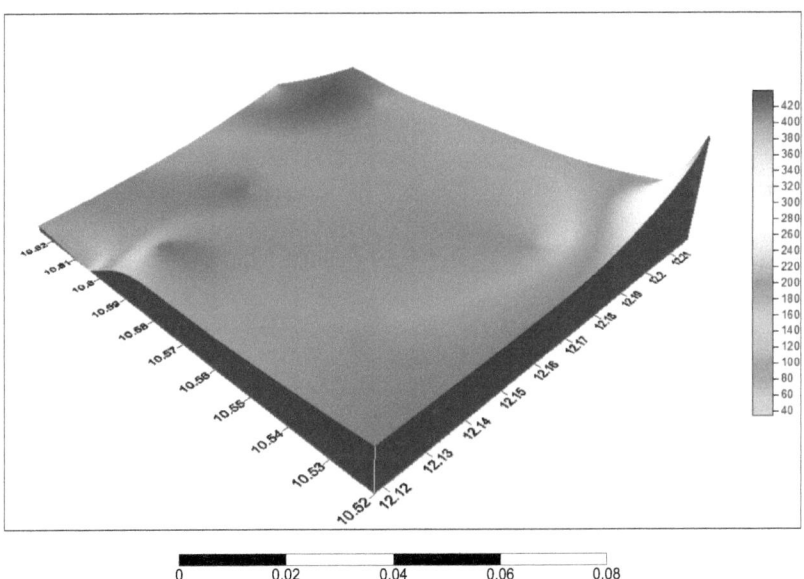

Figure 35: Cr Concentrations in Biu Volcanic Soils in Bar Chat and 3D Surface Map

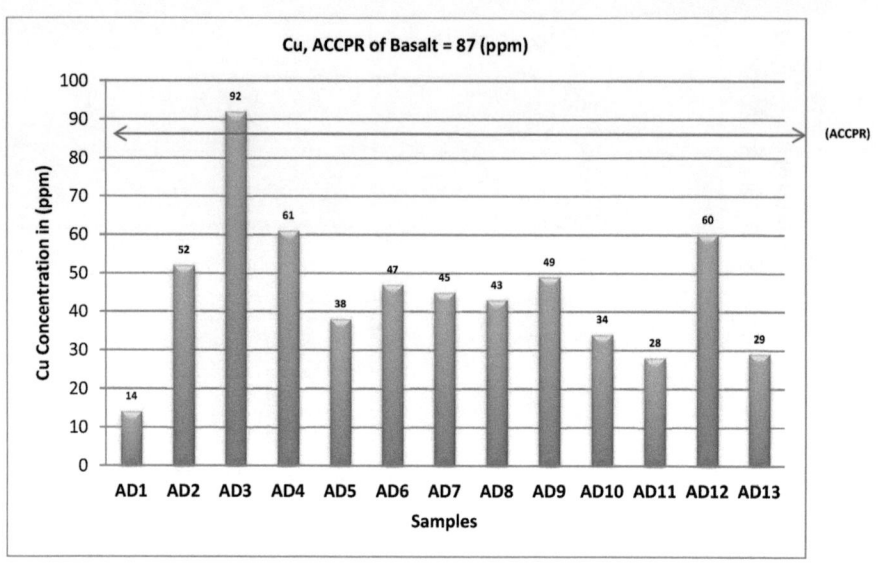
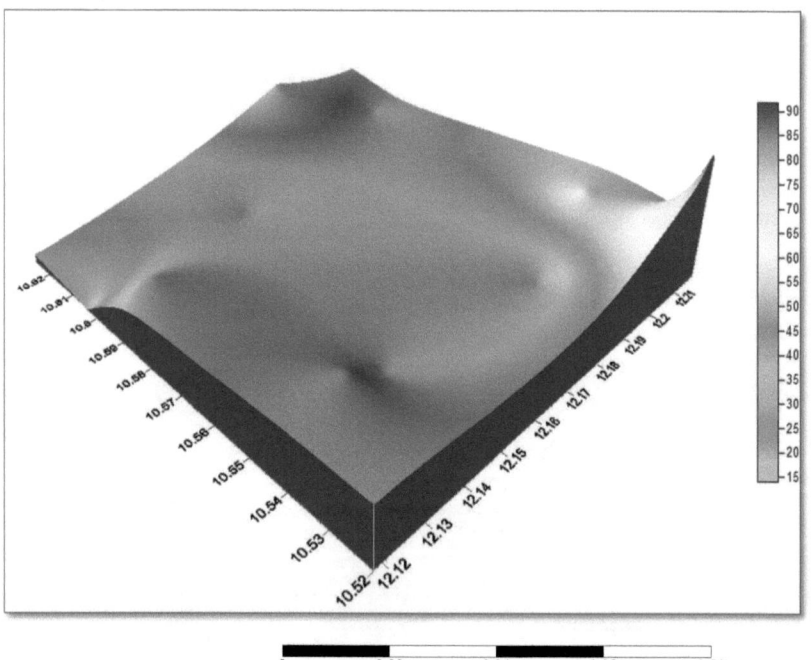

Figure 36: Cu Concentrations in Biu Volcanic Soils in Bar Chat and 3D Surface Map

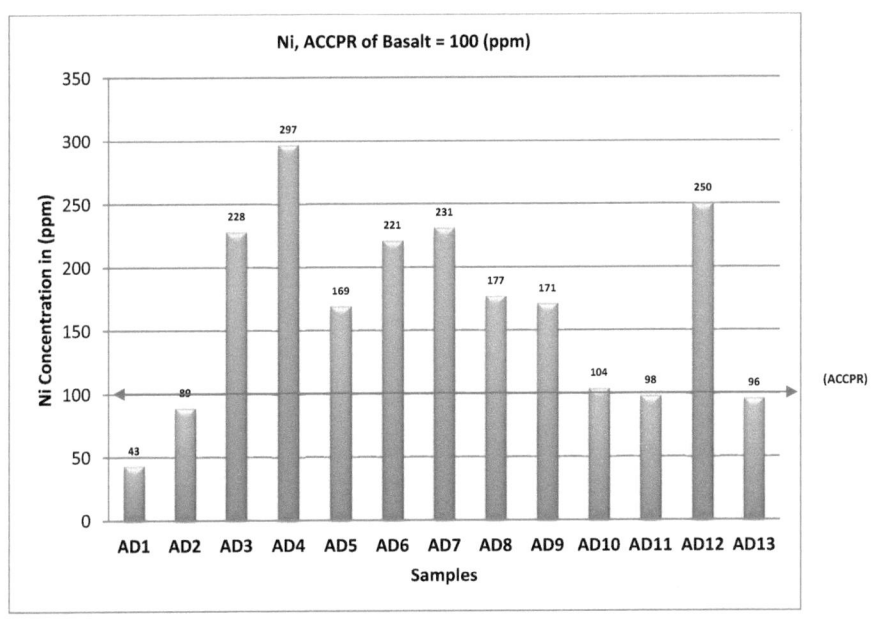
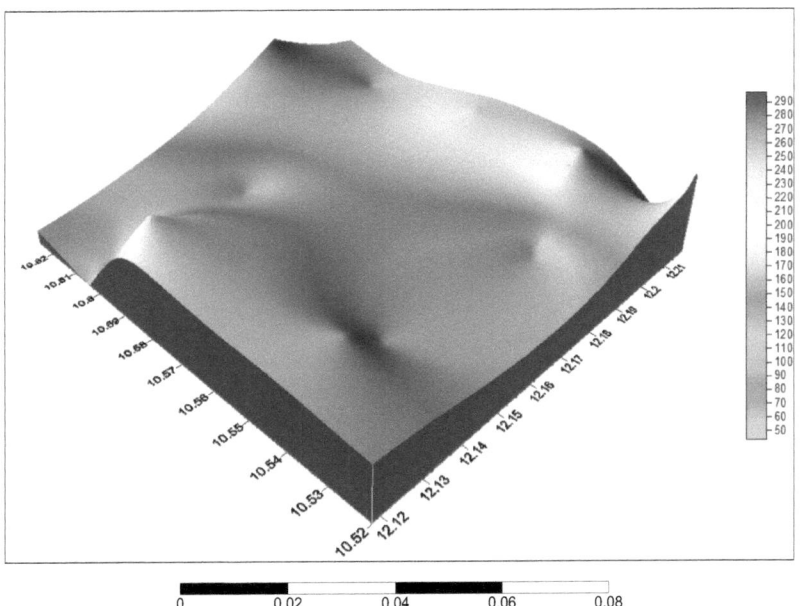

Figure 37: Ni Concentrations in Biu Volcanic Soils in Bar Chat and 3D Surface Map

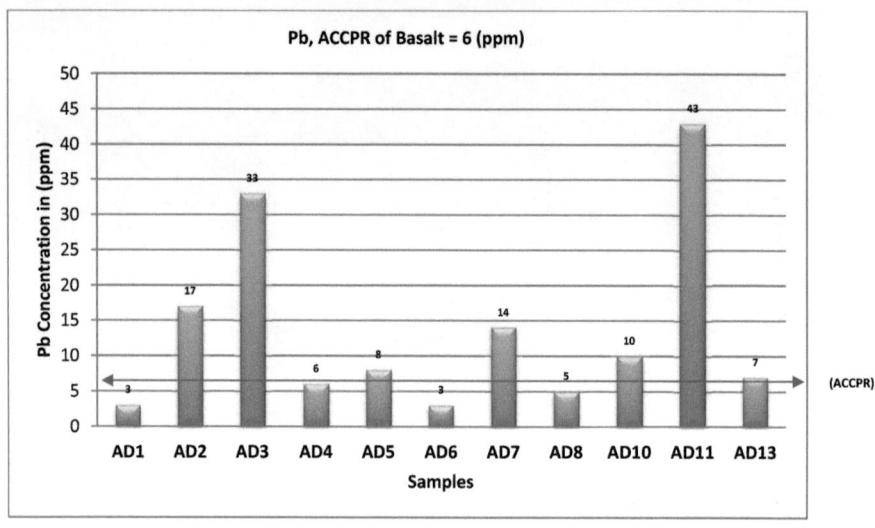
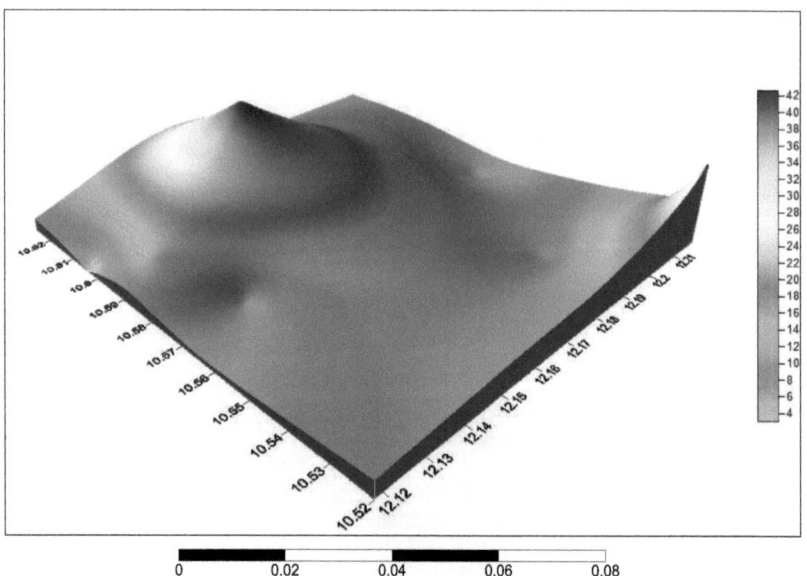

Figure 38: Pb Concentrations in Biu Volcanic Soils in Bar Chat and 3D Surface Map

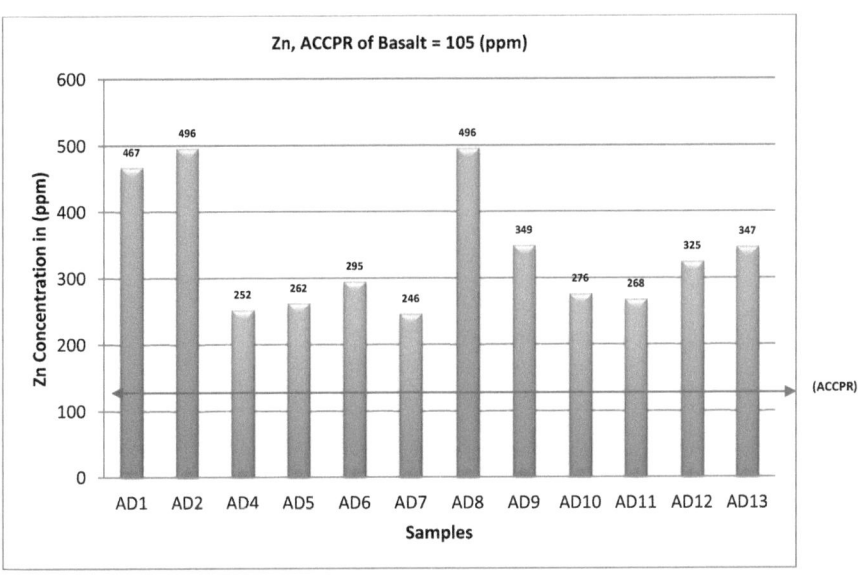
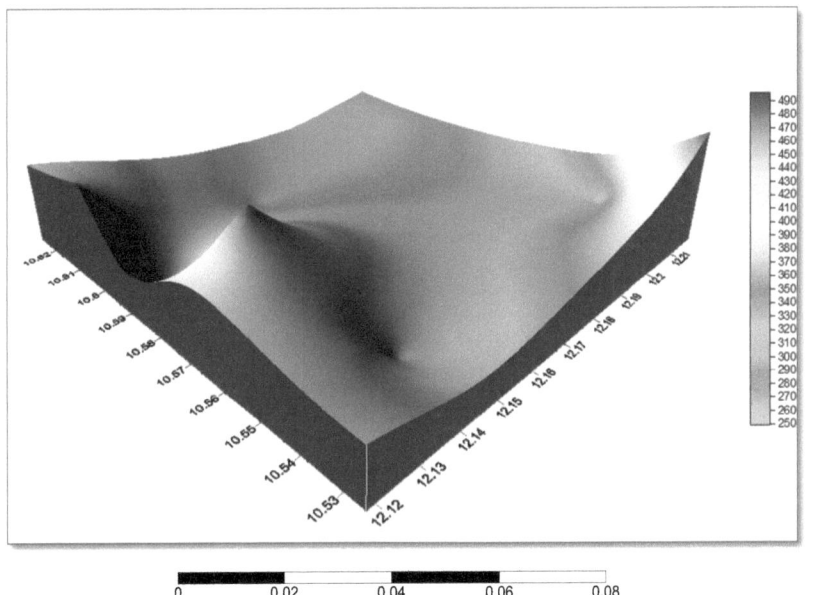

Figure 39: Zn Concentrations in Biu Volcanic Soils in Bar Chat and 3D Surface Map

5.1.2 MAJOR ELEMENTS

Ca, K, Mg, Na, and P

5.1.2.1 Calcium (Ca)

The concentrations of Calcium in surface water (SW) samples collected from various locations in Biu that are above WHO admissible standard are observed in sample (SW2) in Gwarta village with concentration of 340 mg/l and sample (SPW) collected from surface spring water in Yimishika village with concentration of 332 mg/l (fig. 11).

Collected well (W) water samples from various locations in Biu Volcanic Province contained calcium above WHO permissible standard at four different locations. It is observed that (W5) located inside Biu town has concentration of 318 mg/l higher than (W2) from Biu Cotton Ginnery (BCG) with concentration of 300 mg/l and (W18) from Kunar Village with concentration of 283 mg/l while (W14) sampled from Hena village has lowest concentration of 256 mg/l (fig. 11).

From the above result it appears that calcium has concentrations above WHO admissible standard in both surface and well water compared to boreholes with concentrations lower than WHO admissible standard. The concentrations may be due to leaching of high content of Ca containing minerals.

5.1.2.2 Potassium (K)

The observed concentration of potassium in surface water collected from different locations in Biu Volcanic environment are from stream water sample (SW2) from Gwarta Village with concentration of 36 mg/l and surface spring collected from Yimishika water with concentration of 15 mg/l, both the samples are concentrated above WHO admissible standard of 12 mg/l. (fig. 12).

The concentration of potassium occurs only in well two (W2) collected from Biu Cotton Ginnery (BCG), probably due to leaching of rocks containing high contents of K minerals with concentration of 23 mg/l (fig. 12). Wells (5, 13, 14, 16, 17 and 18) all are below WHO admissible standard of 12 mg/l.

The high concentration of potassium in samples SW2, SPW and W2 can be compared with the high concentration of Ca in the in the same samples, indicating that all the sampling points SW2, SPW and W2 are concentrated with K and Ca above WHO admissible standard.

5.1.2.3 Magnesium (Mg)

The concentration of Mg in Boreholes varied from 26-62 mg/l. The highest value of 62 mg/l was observed in sample collected from Waka Biu while the lowest value of 26 mg/l was recorded inside Biu Town (fig. 13).

The well water samples have concentration from 7-171 mg/l. The highest value of 171 mg/l was observed in Biu Cotton Ginnery while the lowest value of 7 mg/l observed in Yimishika village.

The observed concentration of Mg in surface water varied from 7-199 mg/l. The highest concentration of 199 mg/l was recorded in Gwarta while the lowest concentration of 7 mg/l was observed in Tan village (fig. 13).

5.1.2.4 Sodium (Na)

The concentrations of Na in the water samples are below the WHO admissible standard of 200 mg/l. Observed concentration of Na in Boreholes varied from 23-78 mg/l. The highest concentration of 78 mg/l was noted in Biu Town while the lowest concentration of 23 mg/l was observed in Waka Biu (fig. 14).

The observed concentration of Na in well water (W) varied from 2-68 mg/l. The highest concentration of 68 mg/l was observed in Biu Cotton Ginnery (BCG) while the lowest concentration 2 mg/l was collected from Fillin Jirgi Biu.

The concentration of Na in surface water samples (SW) is very low in all the sampled water sources in Biu Volcanic environment. The concentration varied from 0.013-0.071 mg/l. The highest value of 0.071 mg/l was observed in Gwarta while the lowest value of 0.013 mg/l was recorded in Biladega Village.

5.1.3 TRACE ELEMENTS

A total of 15 trace elements (As, Ba, Cd, Co, Cr, Cu, Fe, I, Mn, Mo, Ni, Pb, Sb, Se, Sr and S) were analysed in the water samples.

5.1.3.1 Arsenic (As)

Figure 15 shows total arsenic concentrations in the samples from both sampling areas. It is noted that As is present in all the water bodies at concentration above WHO admissible standard of 0.01 mg/L.

The variation of As in surface water (SW) range from 0.02-0.5 mg/l. The lowest concentration of 0.02 mg/l was observed in Takwa village while the highest concentration of 0.5 was detected in surface spring water in Yimishika village.

Concentration of As varied from 0.01-0.4 mg/l in well water (WW) samples collected from various sites. The lowest concentration of 0.01 mg/l detected from Tila while the highest concentration of 0.4 mg/l was determined from Kunar.

The observed concentration of As in borehole water (BH) samples collected from various sites in Biu ranged from 0.04-0.2 mg/l. The highest concentration of 0.2 mg/l was recorded in Waka Biu while the lowest concentration of 0.04 mg/l was observed in Biu Army Barrack.

5.1.3.2 Barium (Ba)

Figure 16 shows the concentration of barium (Ba) against sample location points. In all the water samples analysed, the concentration is below the WHO admissible standard of 0.3 mg/L.

Concentration of Ba in boreholes (BH) ranges from 0.02-0.03 mg/l while in well (W) varied from 0.01-0.2 mg/l and in surface water (W) the concentration ranges from 0.01-2 mg/l (fig. 16).

5.1.3.3 Cadmium (Cd)

The surface water samples collected at various locations in Biu Volcanic environment with concentration above WHO admissible standard of 0.03 mg/l are surface water (SW2) collected in Gwarta village with concentration of 0.007 mg/l and surface spring (SPW) collected at Yimishika village with concentration value of 0.005 mg/L (fig. 17). The borehole water sample that is concentrated above WHO admissible standard is BH 2 with concentration of 0.04 mg/l while well water sample (W 2) collected at Biu Cotton Ginnery (BCG) has concentration of 0.006 above WHO admissible standard (fig. 17).

5.1.3.4 Chromium (Cr)

Figure 18 shows the concentration of Chromium (Cr) against sample location points. In all the water samples analysed, the concentration is below the WHO admissible standard (0.05 mg/ L). Concentration of Cr in borehole water (BH) varied from 0.001 mg/l in BH 3 and BH 4 (Hema and Biu Town) to 0.01 mg/l in Waka Biu. In well water (W) the concentration of Cr varied from 0.01mg/l in W 6, W 7, W 11 and W 13 (Filin Jirgi, Biladega, Tila and Hena) to 0.03 mg/l in W 18 (Kumar). Variation of Cr concentration in surface water rages from 0.01-0.03 mg/l, the highest value is concentrated in surface spring water SPW sampled at Yimishika village while the lowest value of 0.01mg/l was observed in Waka Biu.

5.1.3.5 Copper (Cu)

The concentration of Cu in all the water samples analysed are below WHO admissible standard of 2 mg/l (fig. 19). The variation of Cu concentration in boreholes (BH) ranges from 0.004-0.02 mg/l while in well (W) the concentration varied from 0.03-0.1 mg/l. In surface water (SW) the concentration varied from 0.004-0.1 mg/l.

5.1.3.6 Iron (Fe)

Figure 20 shows the concentration of Iron (Fe) against sample location points. There is no health concern at concentration level normally observed in drinking-water; taste and appearance are affected below the health based value.

WHO 2003, suggested that concentration above 1.0 mg/l would markedly impair the portability of the water. Concentration value of 1.5 mg/l which is above WHO admissible standard of 1 mg/l appears only in Well 5.

5.1.3.7 Iodine (I)

Available data is inadequate to permit derivation of health-based value and lifetime exposure to Iodine through water disinfection in unlikely (WHO, 2003). For these reasons a guideline value for Iodine has not been established at this time.

The variation of I in boreholes (BH) ranges from 13-28 mg/l while well water (W) concentration varied from 0.2-14 mg/l and in surface water the I concentration ranges from 8-88 mg/l (fig. 21).

5.1.3.8 Manganese (Mn)

The concentration of Mn varied from 0.03- 0.1 mg/l in boreholes (BH) while in well (W) the concentration varied from 0.02-0.7 mg/l. The observed concentration of Mn in stream water (SW) varied from 0.01-0.2 mg/l.

The concentration of Mn from various sources in Biu Volcanic rocks falls below WHO admissible standard of 0.5 mg/l except in W5 from Biu Town with concentration of 0.7 mg/l (fig. 22).

5.1.3.9 Molybdenum (Mo)

Figure 23 shows the concentration of Molybdenum against sample location points. Peaks of Mo concentration above WHO admissible standard (0.07 mg/l) were observed in sample W 18 from well in Kunar and sample SPW from surface spring in Yimishika. Variation of Mo in borehole (BH) ranges from 0.007-0.4 mg/l while the concentration in well (W) varied from 0.002-0.08 mg/l and in surface water (SW) the observed concentration varied from 0.004-0.1 mg/l.

5.1.3.10 Nickel (Ni)

The borehole water sample collected from various sources in Biu Volcanic province has concentration range from 0.001-0.01mg/l lower than WHO admissible standard of 0.02 mg/l (fig. 24). The highest concentration of 0.01 mg/l was collected in (BH 1) from Biu Town while the lowest value of 0.001 mg/l was sampled from (BH 2) from Hena village. Concentration of Ni in well (W) samples varied from 0.01-0.03 mg/l. The highest value of 0.03 mg/l which is above WHO admissible standard of 0.02 mg/l was collected in Biu Town while the lowest value of 0.001 mg/l was sampled from well (W) 12 at Tabra Fulani. The concentration of Ni in surface water of Biu volcanic rock varied from 0.001-0.02 mg/l. The highest concentration of 0.02 mg/l was observed from surface water (SW) 11 collected from Gwarta and the lowest value 0.001 mg/l was observed from the sample collected from Tila village (fig. 24).

5.1.3.11 Lead (Pb)

Lead was detected in all examined ground and surface waters samples of Biu Volcanic Province above WHO admissible standard of 0.01 mg/L except in Wells (W6 and W7) which falls within the standard (fig. 25). The concentration range of Pb in the collected water samples are in concentration range of 0.01-0.4 mg/l. The highest value of 0.4 mg/l collected from surface spring (SPW) in Yimishika village while the lowest value of 0.01

mg/l collected in wells (W6 AND W7) in Filin Jirgi and Biladega respectively. The samples that have high concentrations are surface water samples followed by well water samples with medium concentration. The lowest Pb concentration comes from Boreholes samples. The high concentration in the surface might be due to direct enrichment of the water from the surrounding rocks country rocks. The low concentration in the borehole may be due present of clay which absorb the element in its interstitial space.

5.1.3.12 Antimony (Sb)

Antimony was detected in all analysed water samples from Biu Volcanic Province are above WHO admissible standard of 0.005 mg/l (fig. 26). The concentration of Antimony in Boreholes (BH) range from 0.01mg/l in BH4 (Biu) to 0.03 mg/l in BH1 (Waka) while in Wells (W) the concentration range from 0.01 in W11 (Tila) to 0.05 in W17 (Army Barrack) and in surface water (SW) the concentration range from 0.01mg/l in SW6 (Biladega) to 0.07 mg/l in SW2 (Gwarta).

Stream waters have higher concentration of Antimony than wells with medium concentration; the lowest concentration appears in Boreholes.

5.1.3.13 Selenium (Se)

Figure 27 shows the concentration of Selenium (Sb) against sample location points. The peak of Se concentrations are observed in the entire water samples with concentration higher than WHO admissible standard (0.01 mg/l) except in surface samples SW9 (Takwa) and SW10 (Tan) with concentrations of 0.01 mg/l within WHO admissible standard of (0.01mglL).

The concentration of Se in Boreholes varied from 0.05mg/l in BH3 (Army Barrack) -0.14 mg/l in BH1 (Waka) while in wells the variation is from 0.02 mg/l in W3 and W4 (Yimishika and Gwarta) – 0.28 mg/l in W18 (Kunar) and in Surface Water the concentration varied from 0.01 mg/l in SW9 and SW10 (Takwa and Tan) – 0.44 mg/l in SW2 (Gwarta).

5.2 Soil Sample Analysis Results

Soil samples reflect variations of the geogenic composition of the uppermost layers of the Earth's crust. Trace elements in the soil we cultivate, find their way into the human body directly from the soils and/or underlying bed rock. The natural trace element compositions of rocks and soil may become direct risk of human health and may be the underlying cause of element deficiency and toxicity. The water we drink has a history of percolation through rocks and soils as part of water cycle, and will have leach out chemical elements in solution.

Average Chemical Composition of Parent Rock (ACCPR) of Biu Plateau Basalt (Saidu, 2004), will be use as baseline to understand the distribution and concentration of the elements in volcanic soils of the study area. The result of the analysis is discussed below. A total of 13 soil samples were analysed for major and trace elements.

5.2.1 MAJOR ELEMENTS

CaO, MgO, K_2O, MnO and TiO

These major elements are concentrated below average major elements composition of Biu Plateau Basalt (fig 28, 29, 30, 31 and 33). Only Fe_2O_3 (fig. 29) present values higher than Average Chemical Composition of Parent Rock (ACCPR) of 10.71 % in almost all the samples collected i.e. (AD2, AD2, AD3, AD4, AD5, AD6, AD7, AD8, AD9, AD10, AD11, AD12 and AD13) with concentrations of 49.81, 85, 28.41, 19.33, 29.87, 34.71, 14.26, 16.29, 30.63, 23.1, 42, and 13 % respectively.

5.2.2 TRACE ELEMENTS

Co, Cr, Cu, Ni, Pb and Zn

5.2.2.1 COBALT

The concentration of Co in figure 34 varied from 13 – 111 ppm. Three peaks of cobalt concentrations above average composition of Biu Basalt (60 ppm) are noted in sample AD3 collected in Gwarta and AD7 collected from Biladega with concentrations of 111, 111, 84 ppm respectively, the lowest concentration of 13 ppm was sampled from Biu Cotton Ginnery (BCG).

5.2.2.2 CHROMIUM

Figure 35 shows the concentrations of Cr which varied from 34-441ppm. Three samples (AD2, AD3 and AD4) with concentrations of 230, 441 and 247 ppm have high anomaly above average chemical composition of Biu Basalt (170 ppm). The highest concentration represented by sample AD3 was collected from Gwarta followed by sample AD4 collected from Waka, the lowest concentration sample AD2 was collected from Yimishi-ka.

5.2.2.3 COPPER

Detected concentration of Co above average composition of Biu Basalt (87 ppm) was observed in sample AD3 collected from Waka with concentration of 92 ppm (fig. 36). The lowest Cu concentration was detected in sample AD1 from BCG.

5.2.2.4 NICKEL

Figure 37 displays the concentrations of Ni in the soil samples collected from Biu Volcanic province. The peaks of Ni concentrations were observed in samples (AD3, AD4, AD5, AD6, AD7, AD8, AD9 and AD12) in Gwarta, Waka, Filin Jirgi, Biladega, Tila and Kunar with concentrations of 228, 297, 169, 221, 231, 177, 171, 250 ppm respectively. The variations range from 43-297 ppm, the highest concentration was observed in AD4 sampled from Waka, while the lowest concentration was detected in sample AD1 collected from Biu Cotton Ginnery (BCG).

5.2.2.5 LEAD

Concentration of Pb in collected soil samples of Biu Volcanic Province varied from 3-43 ppm (fig.38). High concentration above average composition of Biu Basalt (6 ppm) are noted in samples collected in Yimishika (AD2), Gwarta (AD3), Waka (AD5), Biladega (AD7), Tila (AD10), Malan (AD11) and Takwa (AD13) with concentrations of 17, 33, 8, 14, 10, 43 and 7 ppm respectively.

5.2.2.6 ZINC

Figure 39 shows the concentrations of Zinc in Biu Volcanic soils. Peaks of Zinc concentrations are observed in all collected samples above average composition of Biu Basalt (105 ppm). The variation of Zn concentrations is from 246 – 496 ppm. The highest value of 496 ppm was observed in sample AD2 and AD8 from Yimishika and Biladega respectively while the lowest of 246 ppm was recorded in sample AD7 from Biladega.

5.3 DISCUSSION OF RESULT

5.3.1 MAJOR ELEMENTS IN SOIL AND WATER SAMPLES

5.3.1.1 SOIL

MgO, CaO, TiO_2, K_2O and Fe_2O_3

These elements are highly enhanced in the weathered volcanic soils. This is clearly observed when compared with Parent Rock, for example the concentrations of MgO, CaO, TiO_2 and K_2O are 10.6, 9.95, 2.23, 2.07 and 10.71 % in parent rock compared with concentrations of 11, 22, 10.3, 10.2 and 30 % respectively in weathered volcanic soils (table 9). Fe_2O_3 shows high anomaly in the soils samples (49.81, 85, 28.41, 19.33, 29.87, 34.71, 14.26, 16.29, 30.63, 23.1, 42, and 13 % respectively, Table 7), because it is easily mobilized during weathering process resulting in elevated concentration compared to parent rock with concentration of 10.7%. However, near absence in water could be its inability to remain in solution.

The average concentrations of K_2O, CaO, MgO, Fe_2O_3 and TiO_2 in soils are 173, 126, 123, 46 and 20% higher than average concentration in groundwater (4.5, 94, 45.5, 0.06 and 0.002 ppm) and surface water (39, 38, 29, 0.9 and bdl), Table 10.

These show that the major elements are mobilized from the leaching of the volcanic soils leading to their concentration in surface and groundwater.

5.3.1.2   WATER

Ca, K, Mg, and Na

The major cations Mg (7-199 mg/l) and Na (2.4-78 mg/l) are within the minimum permissible drinking water level, however, Calcium shows elevated concentrations in wells, surface water and spring water of 300, 318, 255, 283, 240 and 332 mg/l respectively, the concentration may be due to leaching of Ca rich minerals in the volcanic soils (Figure 11). Potassium shows high enrichment values of 15.37, 22.51 and 36.16 mg/ in spring water (SPW), surface water (SW) and well (W).

Total hardness is normally expressed as total concentration of Ca^{2+} and Mg^{2+} in milligram per litre $CaCO_3$. The total hardness values of the waters in Biu volcanic Province ranges from 70 to 1666 mg/l as shown in appendix 2, indicating that the water from boreholes, wells and surface waters are moderate to very hard, therefore, not suitable for both drinking, washing and bathing.

5.3.2 TRACE ELEMENTS IN SOIL AND WATER SAMPLES

5.3.2.1 SOIL

As, Cd, Co, Cr, Cu, Ni, Pb, Zn, Sb, and Se

These elements As, Se, Cd and Sb are below detection limit in the weathered volcanic soils; this could be due to their high mobility resulting in their depletion and concentration into the water sources.

When this study is compared with parent rock, Co recorded high anomaly of 111 and 84, while, Cr has three anomalies of 441, 247 and 230 ppm and Cu is nearly depleted and shows concentration of 92 ppm in one sample only.

Ni shows high abundance, some samples shows more than two fold concentration. Pb shows elevated concentration compared to parent rock due to its immobility. Zn present high anomaly greater than parent rock in all the samples, the concentration ranges from 246 – 496 ppm, this could be its ability to be absorb by clay particles.

5.3.2.2 WATER

As, Se, Sb, Pb and Cd

These elements are found in appreciable amount far above WHO admissible standard.

5.3.2.2.1 Arsenic

Arsenic is completely depleted in the volcanic soils but present high values of 0.48, 0.42, 0.39 and 0.32 mg/l in surface, well and spring water respectively. These values are low when compared the parent rock, Table 9.

The average concentration of As in surface water is 0.49mg/l higher than groundwater with concentration of 0.12mg/l, (Table10).

5.3.2.2.2 Selenium

Se concentration is very high with values of 0.44, 0.43, 0.28, 0.27, 0.26, and 0.25 mg/l above WHO value of safe drinking water of 0.01mg/l in surface, well and spring water, but completely leached away from the soils and parent rock due to its high mobility resulting in elevated concentration in surface and well waters.

The average concentration of Se in surface water is 0.48mg/l higher than groundwater with concentration of 11mg/l (table10).

5.3.2.2.3 Antimony

Sb presents high value of 0.071, 0.047, 0.041 and 0.034 mg/l in surface, well, spring and borehole waters while it is depleted soils and parent rock. These values are far above admissible value of safe drinking water.

The average concentration of Sb in surface water is 0.03 higher than average concentration in groundwater, Table 10.

5.3.2.2.4 Lead

Elevated concentration of lead with values of 0.37, 0.35, 0.31, 0.21, 0.21 and 0.17 mg/l were recorded in surface water and borehole waters respectively. These values are low compared to concentrations in volcanic soils 14ppm and 6ppm in parent rock, table 9.

Based on the computed average values the concentration of lead in the surface water is 0.5 mg/l compared to 0.1 mg/l in groundwater, table 10.

5.3.2.2.5 Cadmium

Cd In the study area has high values of 0.05, 0.007, 0.006 and 0.004 mg/l respectively above WHO admissible standard of 0.003 mg/l in spring, surface, well, and borehole waters. The concentration is below detection limit in parent rock but concentrated in volcanic soils, indicating that its mobility is high, table 9.

The average concentration in surface and groundwater are 0.01 and 0.003 lower than average concentration in parent rock, table 10.

From the discussion above, these elements: MgO, K_2O, CaO, TiO_2 and Fe_2O_3 are concentrated in the volcanic soils higher than the parent rock, leading to their concentration in the weathering profile. This shows that these major elements are mobilized from the leaching of the volcanic soils leading to their concentrations in surface and groundwaters.

Trace elements CO, Ni, Cu, Zn and Pb are distributed in all the soils because of their immobility. V which is relatively mobile is concentrated in only three samples.

Owing to its low mobility Pb is concentrated in all the soil samples, its dissolved fraction which is toxic to Humans and Animals is distributed across the water bodies. These trace elements: As, Se, Mo, Cd, Sb and TI are depleted in all the soils samples because of their high mobility resulting in their concentration in all the water bodies. It is therefore, evident that the source of these toxic elements in waters of the Biu Volcanic Province could be geogenic.

Finally, these trace elements (As, Pb, Sb and Se) shows elevated concentrations in the surface and ground water far above WHO's admissible drinking water standard, this high level will pose environmental and health hazard to the inhabitants of the study area.

5.4 Trace Element Exposure and Human Health

5.4.1 Introduction

The chemical, physical and biological characteristics of water determine its usefulness for industry, agriculture or the home. Drinking water standards adopted by WHO are based on two criteria:

- Presence of objectionable substance such as odour and taste
- Presence of substance with adverse physiological effects such as trace elements. The limits of ions such as arsenic, selenium, chromium, lead etc, are based on adverse physiological effects.

This research focused particularly on the effects of these trace elements to human health at higher or lower level in drinking water of Biu Volcanic Province.

5.4.2 Trace Element Exposure

This study discover the concentration of arsenic form 0.1 to 0.45, the high concentration will be hazardous to the inhabitants due to long time exposure through ingestion of food and water. According to Maloney, 1996 and Smedley and Kinniburgh, 2002, the overexposure to this element can cause various diseases such as cancer (skin, lung, bladder, and kidney), hair loss and nails deformity. This is in conformity with the present study in which some of the inhabitant's shows manifestation of some of these diseases which may be linked to arsenic toxicity.

According to Johnson et al, 2000, the most frequently reported symptoms of selenosis are hair loss, nail brittleness and other symptoms include gastrointestinal disturbances, skin rashes, garlic breath odor, fatigue, irritability, and nervous system abnormalities. In the study area some of the inhabitants are affected by nails brittleness, skin rashes, hair loss which may liken to selenium exposure.

Antimony concentration ranges from 0.01 to 0.03mgl and lead from 0.1 to 0.5, this high concentration will cause overexposure, through the ingestion of food and water.

The effect of antimony include "antimony spots" a form of dermatitis, and later respiratory, pulmonary and heart effects were noted and cancer was suspected (Fowler et al, 1991). Lead is toxic to many organs and tissues including the heart, bones, intestines, kidneys, and reproductive and nervous systems (Grant, 2009). It interferes with the development of the nervous system and is therefore particularly toxic to children, causing potentially permanent learning and behaviour disorders.

The toxic effects of antimony and lead were not encountered. Medical doctors and allied professionals are to be engage in order to unravel the toxicity of the elements to the inhabitants of the volcanic environment.

5.5 Trace Elements and Human Health Impact

The presence of Potentially Harmful Elements (PHEs) (As, Se, Sb, and Pb) in the spring water, stream water and well water in alarming concentrations far above WHO admissible standard in the study area could have adverse health hazards.

Because of the protocols observed in finding out the effect of trace element due to over exposure in Biu Volcanic Province, a case study is pick for proper studies that is Yimirshika Town in which the inhabitants relies on their Spring Water for drinking and other domestic purposes. This spring water in which the Town derives its name contain unsafe levels of dissolved Potentially Harmful Elements (PHEs) (As, Se, Sb, and Pb), these findings indicate that people living in Yimirshika Town are over-exposed to these toxic metals through the ingestion of water and food.

The study discovered that due to long time exposure, few of the inhabitants show manifestations of nail deformity (nail thickening and brittleness), and hyper-pigmentation of the skin and hand palms. Others present various forms of skin diseases (especially skin growth) which all could be attributed to exposure to As and Se toxicity. Plate 5-16 shows various nails deformation and skin problems.

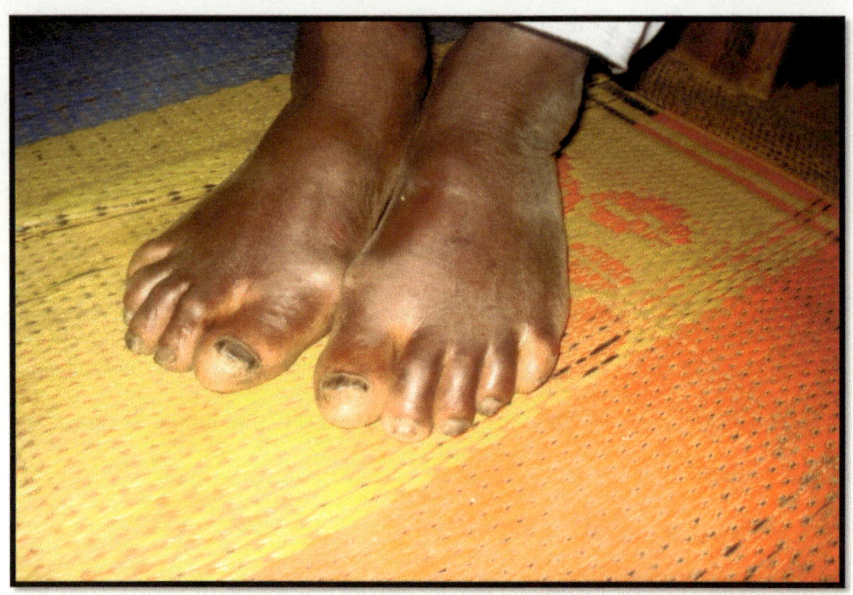

Plate 5: Deformed nails due to As and Se toxicity (12°14'41.4" 10°31'52.2")

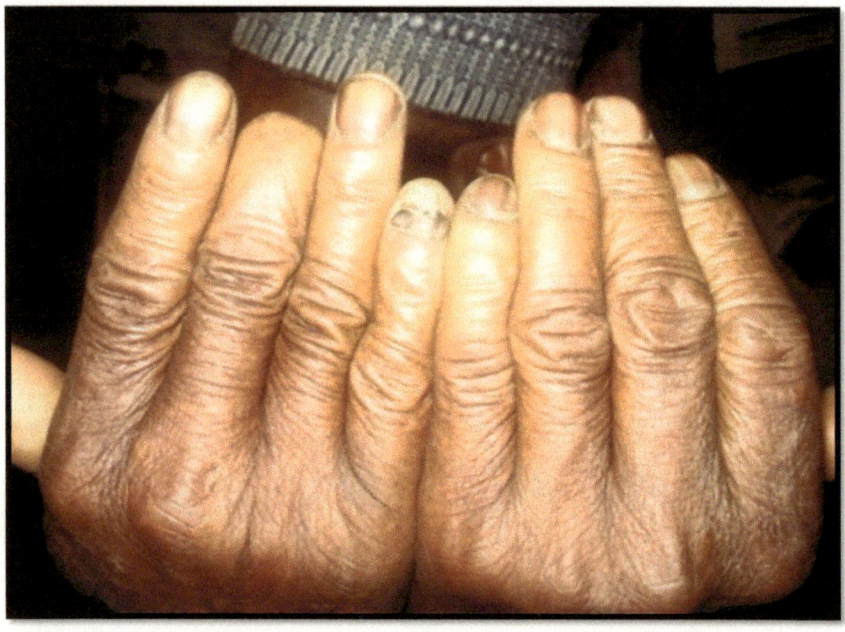

Plate 6: Deformed nails due to As and Se toxicity (12°14'41.4" 10°31'52.2")

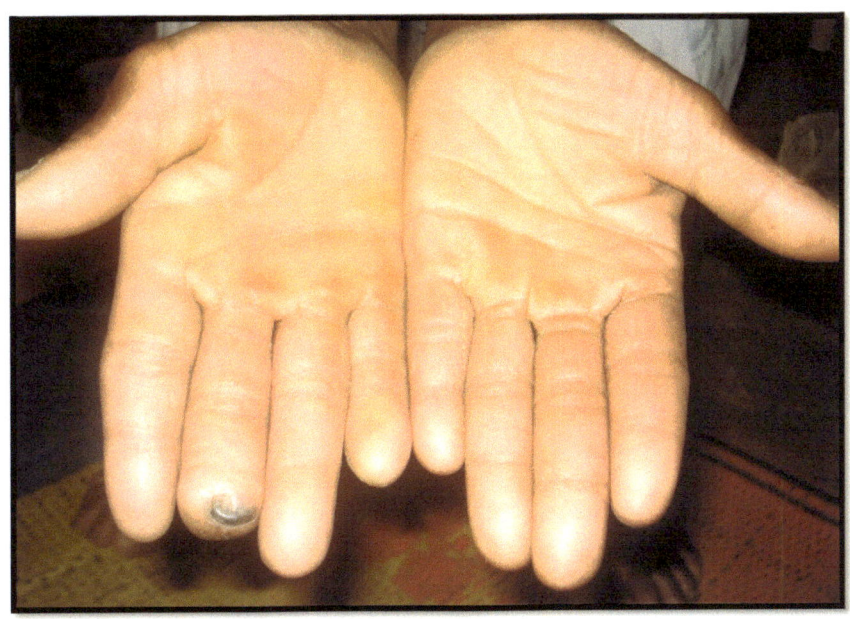

Plate 7: Deformed nails due to As and Se toxicity (12°14'41.4" 10°31'52.2")

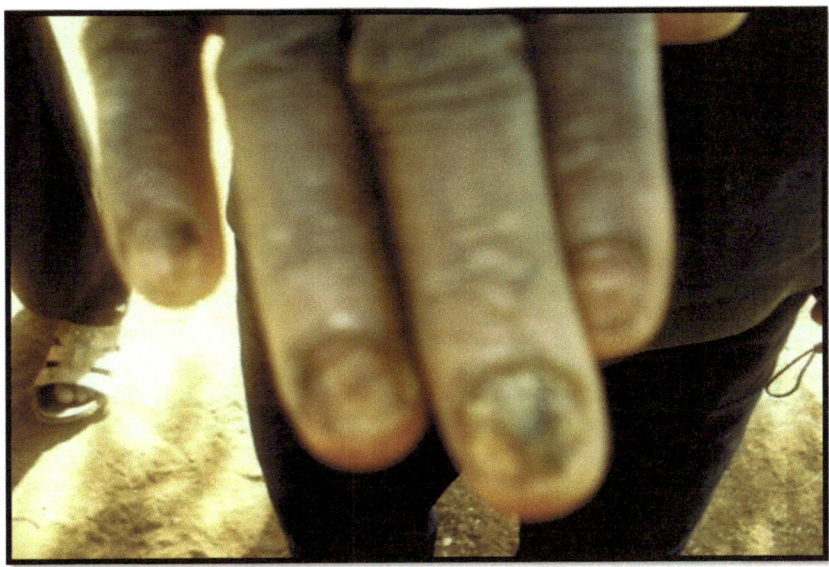

Plate 8: Deformed nails due to As and Se toxicity (12°14'41.4" 10°31'52.2")

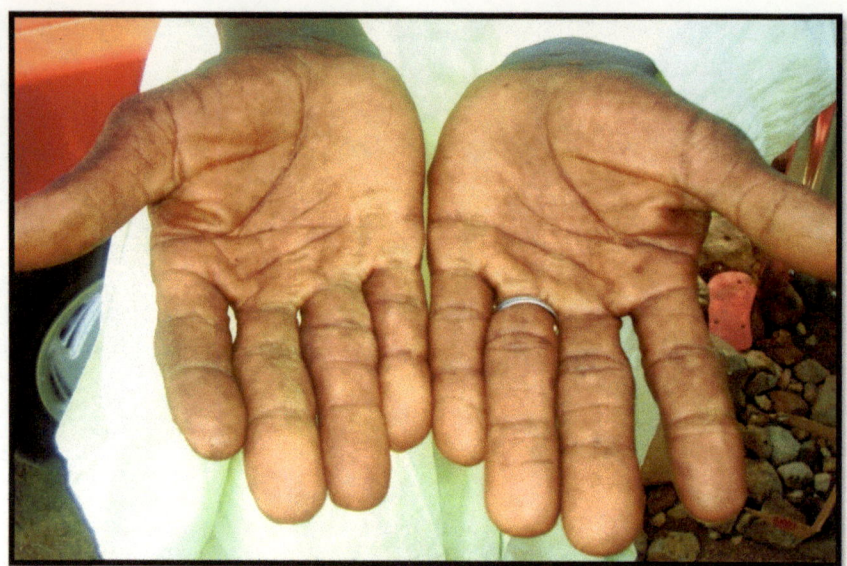

Plate 9: Hyper pigmentation on palms due to As and Se toxicity (12° 14'41.4" 10° 31'52.2")

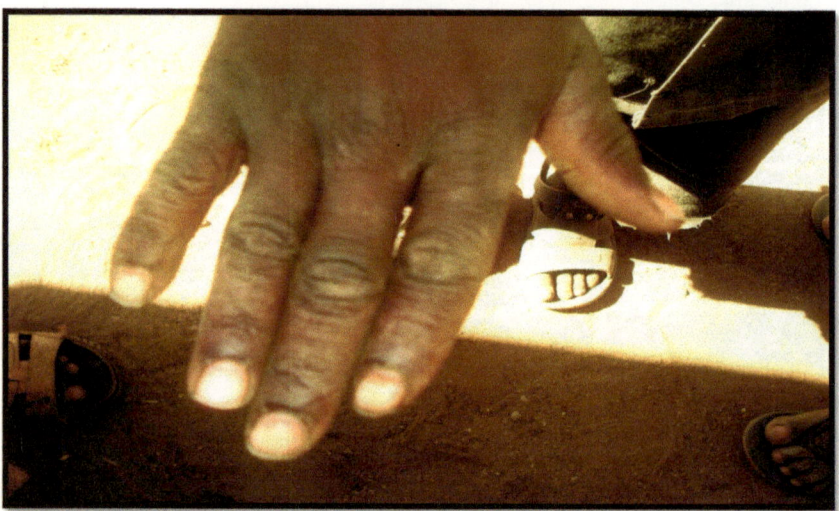

Plate 10: Roughness of the skin and nails brittleness of (12 years Old boy) due to As and Se toxicity (12° 14'41.4" 10° 31'52.2")

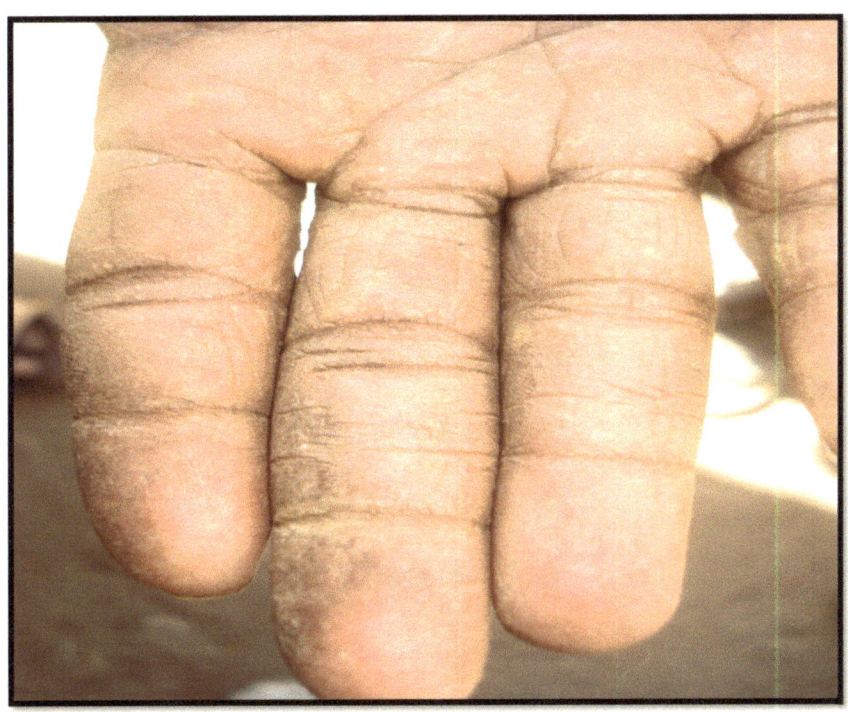

Plate 11: Patches and roughness on palms due to As and Se toxicity (12°14'41.4" 10°31'52.2")

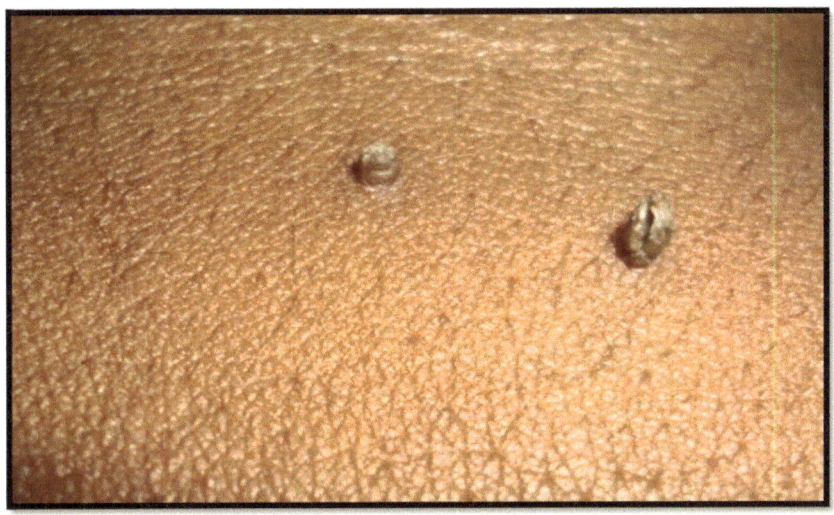

Plate 12: Growth on skin due to As and Se toxicity (12°14'41.4" 10°31'52.2")

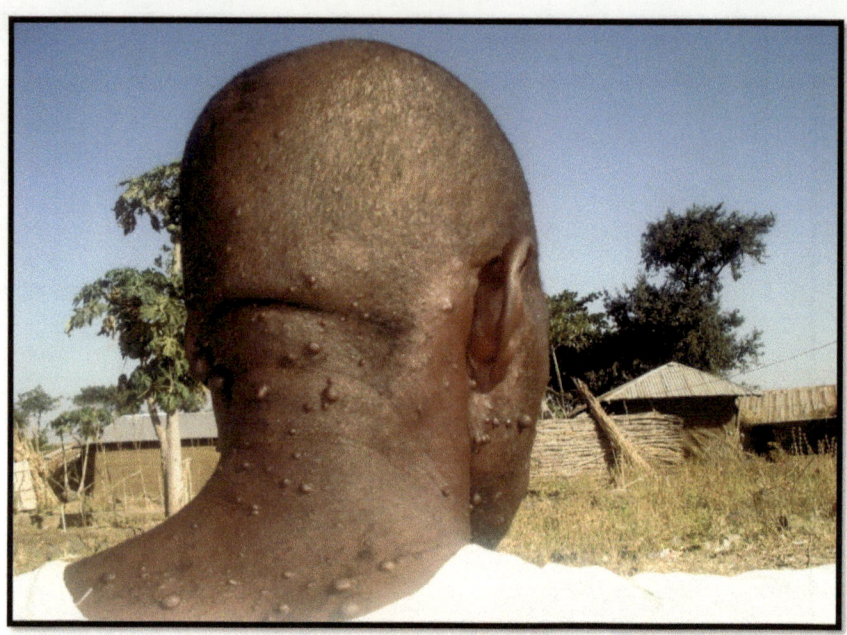

Plate 13: Growth on skin due to As and Se toxicity (12°14'41.4" 10°31'52.2")

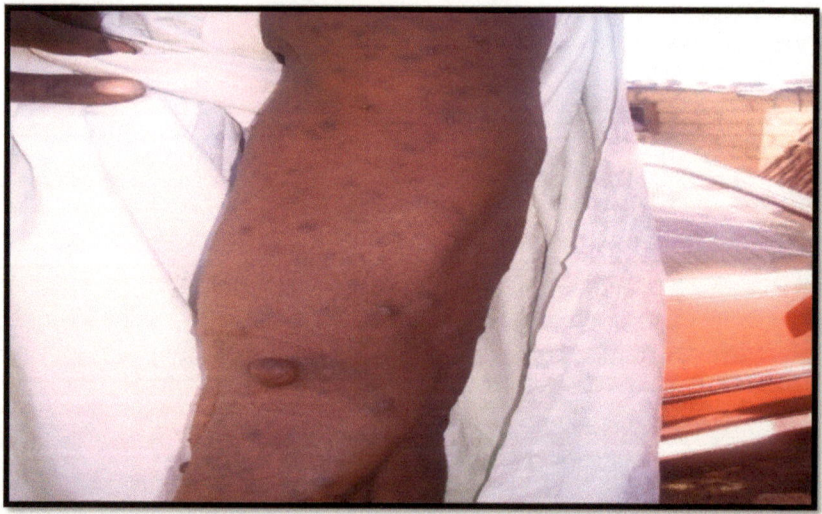

Plate 14: Growth on the skin and rashes due to As and Se toxicity (12°14'41.4" 10°31'52.2")

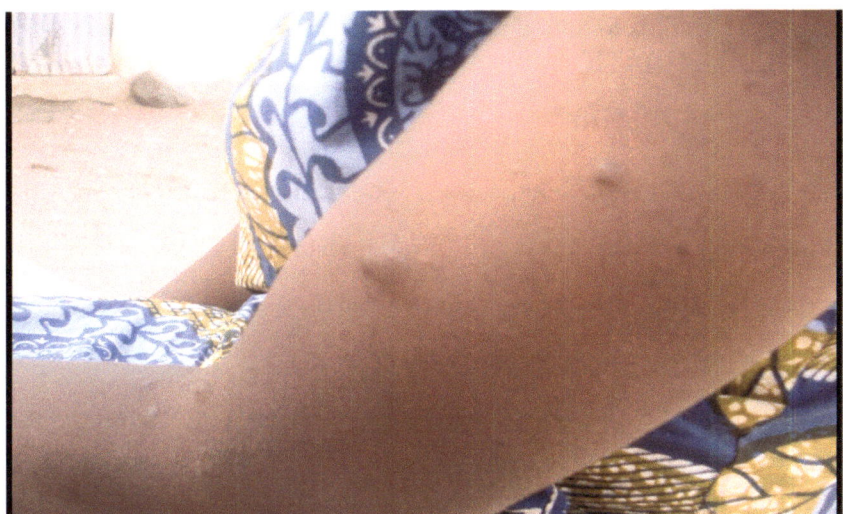

Plate15: Growth on skin due to As and Se toxicity (12°14'41.4" 10°31'52.2")

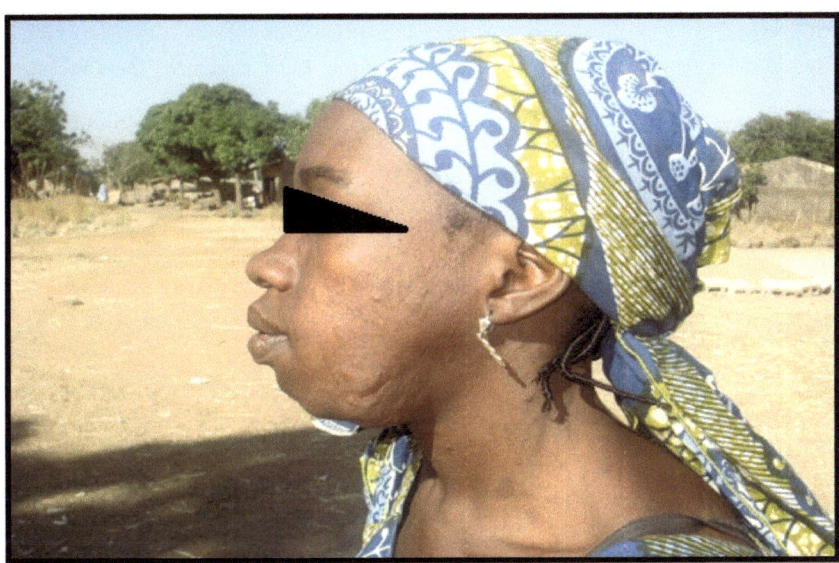

Plate 16: Growth on the jaw due to As and Se toxicity (12°14'41.4" 10°31'52.2")

CHAPTER SIX: SUMMARY, CONCLUSION / RECOMMENDATION

6.1 SUMMARY

Biu Plateau is situated on the structural and topographic divide between the Benue and Chad sedimentary basins. The structural divide is a broad E-W ridge or swell of Basement, which extends to the western edge of the Biu Plateau. The two basins are divided by the Zambuk ridge to the west (Carter, et al., 1963).

The basalt of the Biu Plateau mainly overlies Basement rocks. These are predominantly granites, granite-gneiss and Fayalite-quartz, Monzonite, Bauchites (near Wandali at the SW margin of the plateau), hypersthenes diorite, volcanic and sub volcanic rocks of the Burashika group (Turner, 1978). To the west and north, Basalt of the Biu Plateau has spread over cretaceous sediments, mainly the arkosic Bima sandstone. These rocks are folded, with axes to the SW of the Plateau trending NE-SW, the structures extending into the basement rocks as NE-SW faults (Turner, 1978).

Geochemical analysis of the volcanic soil revealed the complete leaching of the major elements (Fe_2O_3, CaO, K_2O, MgO, MnO, and $TiO2$) from the surface soil probably into water sources.

The absence in the soil profile and the extremely higher values of Potentially Harmful Elements (PHEs) (As, Se, Sb, and Pb) in the spring and stream water as opposed to the lower values in the wells and borehole water suggest their extreme solubility, direct leaching and transportation of these elements from the surrounding rocks into the surface water or from anthropogenic source resulting from agricultural practices like application of fertilizers and pesticides or herbicides.

6.2 CONCLUSION

The present study revealed that both surface and ground water are contaminated by As, Pb, Sb and Se. The study discovered that due to overexposure of these toxic elements some people are affected with problem of diabetes, loss of hearing, hair loss, deformed nails and various skin problems like: rashes, abnormal growth, skin lesion and roughness, could be attributed to exposure to As or Se toxicity based on the quoted literature.

Comparison of physical parameters falls within the maximum permissible limit of the World Health Organization (2008).

The research work could serve as reference for future Groundwater research, Environmental/Medical Geology and Mineral Exploration Studies and also, the result may allow Scientists, policy makers and voluntary organisation to initiate programs to assist the area's most affected by the toxic metals documented in this study by devising a simple means of toxic elements removal which is going to be affordable to the inhabitants and also by providing safe drinking water through drilling of boreholes in areas not affected by toxic elements contamination.

6.3 RECOMMENDATIONS

Based on the study conducted in the Biu Volcanic province the following recommendations are made:

- Further investigation of Arsenic speciation should be carried out since Arsenic occurs in natural waters as in organic (As III) which is more toxic or organic (As V) which is less toxic.
- There is need to create public awareness regarding the health risk associated with these toxic elements.

- There is also need to understand the source of contamination of the toxic metals, which could either be geogenic or anthropogenic, in addition it is also recommended to study their exposure and bioavailability.
- Further studies should include microbial investigation and heavy metal plus isotopic compositions, so as to ascertain other quality parameters and hence prescription of necessary measure.
- There is need for the development of collaborative research (with the medics and other scientists) to increase our understanding of the link between the distribution of some of these trace elements and associated health problems in the study area. Such knowledge is essential for the control and management of these health problems.

REFERENCES CITED

Araya, O., Wittwer, F., Villa, A., and Ducom, C. (1990). Bovine fluorosis following volcanic activity in the southern Andes, Vet. Rec., 126, p. 641-642.

Baxter, P.J., Buabron, J.C., and Coutinho, R. (1999). Health hazards and disaster potential of ground gas emissions at Furnas volcano, São Miguel, Azores, J. Volcanol. Geotherm. Res., 92, 95-106.

Carter, J. D., Baber, W. D., and Tait, E.A. (1963). The Geology of part of Adamawa, Bauchi and Borno provinces in North Eastern Nigeria. Bull. Geol. Survey, Nigeria. Bull. No. 30. p. 13-80.

Casadevall, T.J., and Lockwood, J.P. (1995). Active volcanoes near Goma, Zaire: hazard to residents and refugees, Bull. Volcanol., 57, 275-277.

Cronin, S.J., Neall, V.E., Lecointre, J.A., Hedley, M.J., and Loganathan, P. (2002). Environmental hazards of fluoride in volcanic ash: a case from Ruapehu volcano, New Zealand, J. Volcanol. Geotherm. Res., 121, 271-291.

Du preez, J.W. (1949). The Geology and Hydrology of Biu Fivision. Geol. Surv. Nigeria. Unpub. Rept. No. 751, Geological Survey of Nigeria.

Du preez, J.W., and Barber, D. F. M. (1965). The Distribution and Chemical Quality of Groundwater in Northern Nigeria. Bull. Geol. Surv. Nigeria, 36, 93p.

Falconer, J. D. (1911). The Geology and Geography of Northern Nigeria.

Fordyce, F. M., Johnson, C. C., Navaratne, U. R. B., Appleton, J. D., and Dissanayake, C. B. (1998). Studies of Selenium Geochemistry and Distribution in Relation to Iodine Deficiency Disorders in Sri Lanka. British Geological Survey Overseas Geology Series Technical Report WC/98/28.

Fowler, B. A., and Goering, P. L. (1991). Antimony. In: Merian E, ed. Metals and their compounds in the environment: occurrence, analysis, and biological relevance. Weinheim, VCH, pp. 743–750.

Grant, L.D. (2009). "Lead and compounds". In Lippmann, M. Environmental Toxicants: Human Exposures and Their Health Effects, 3rd edition. Wiley-Interscience. ISBN 0471793353.

Hardisson, A., Rodriguez, M.I., Burgos, A., Diaz Flores, L., Gutierrez, R. and Varela, H. (2001). Fluoride levels in publicly supplied and bottled drinking water in the island of Tenerife, Spain, Bull. Environ. Contam. Toxicol., 67, P.163-170.

Heikens, A., Sumarti, S., van Bergen, M., Widianarko, B., and Fokkert, L. (2005). Van Leeuwen, K., Seinen, W., The impact of the hyperacid Ijen Crater Lake: risks of excess fluoride to human health, Sci. Total Environ., 346, P. 56-69,

Hurtado, R., Gardea-Torresdey, J., and Tiemann, K.J. (2000). Fluoride occurrence in tap water at "Los Altos de Jalisco" in the central Mexico region, Proc. of the 2000 Conf. on Hazardous Waste Research, P. 211-219.

Johnson, C. C., Green, K. A., and Liu, X. (2000). Selenium distribution in the local environment of selected villages of the Keshan Disease belt, Zhangjiakou District, Hebei Province, People's Republic of China. Applied Geochemistry. Vol. 15. No. 3.

Johnson, A.H., and Reynolds, R.C. Jr. (1977). Chemical character of head-water streams in Vermont and New Hampshire. Wat. Resour. Res. 13(2), 469-473.

Kawahara, S., Odontological. (1971). observations of Mt. Aso-volcano disease, Fluoride, 4, P.172-175.

Kloos, H., and Tekle Haimanot, R. (1999). Distribution of fluoride and fluorosis in Ethiopia and prospects for control, Tropical Medicine and International Health, 4, P. 355-364.

Lar, U. A., and Tejan, A, B. (2008). Highligts of some environmental problems of geomedical significance in Nigeria. Journal of Environmental Geochemistry and Health. 30:383-389.

Lar, U.A. (2009). Trace Element and Health; Some Case Studies from Nigeria. Nigerian Mapping Technical Assistance Programme, Kaduna (2009).

Maloney, M. (1996). Arsenic in dermatology. Dermatol Surg; 22:301-304.

Morettini, L., and Ciranni, R. (2000). Herculaneum – other mysteries unearthed, C.N.R., Bologna.

Mungoma, S. (1990). The alkaline, saline lakes of Uganda: a review. Hydrobiologia, 208, P. 75-80.

Nogawa, Koji., Kobayashi, E., Okubo, Y., and Suwazono, Y. (2004). "Environmental cadmium exposure, adverse effects, and preventative measures in Japan". Biometals 17 (5): 581–587.

Nanyaro, J.T., Aswathanarayana, U., Mungure, J.S., and Lahermo, P.W. (1984). A geochemical model for the abnormal fluoride concentration in waters in parts of northern Tanzania, J. African Earth Sci., 2, P. 129–140.

Nyaora Moturi, W.K., Tole, M.P., and Davies, T.C. (2002). The contribution of drinking water towards dental fluorosis: a case study of Njoro Division, Nakuru District, Kenya, Environ. Geochem. Health, 24, P. 123-130.

Oppenheimer, C. (2003). "Climatic, environmental and human consequences of the largest known historic eruption: Tambora volcano (Indonesia) 1815". Progress in Physical Geography 27 (2): 230-259.

Pekdeger, A., Özgür, N., and Schneider, H.J. (1992). Hydrogeochemistry of fluoride in shallow aqueous systems of the Gölcük area, SW Turkey, In: Kharaka, Y.K.,Maest, A.S. (eds) Proc. 7th Intern. Symp. on Water Rock Interaction, Utah, P. 821-824.

Saidu, Y. (2004). Geochemical characteristic and evolution of the Biu Plateau basalt. MSc. Thesis UNPBL. Geol. and Min. Dept. Unijos. PP. 4 and 53.

Selinus, O. (2004). Medical Geology: an emerging specialty. Terrae. Vol. 1, No. 1, pp. 8.

Smedley, P. L.,and Kinniburgh, D. G. (2002). "A review of the source, behaviour and distribution of arsenic in natural waters". Applied Geochemistry 17 (5): 517–568.

Soto-Rojas, A.E., Ureña-Cirett, J.L., and Martínez-Mier, E.A. (2004). A review of the prevalence of dental fluorosis in Mexico, Pan Am. J. Public Health, 15, P. 9-17.

Spencer, J.E. (2000). Arsenic in Ground Water. Arizona Geology: Arizona Geological Survey, 30(3), 1-4.

Turner, D. C. (1978). Volcanoes of the Biu Basalts, North eastern Nigeria. Jour. Min. Geol. Vol. 15. PP. 49-62

Ullrey, D. E. (1981). Selenium in the Soil-Plant-Food Chain. In: Selenium in Biology and Medicine, eds. Julian E. Spallholz, John L. Martin, and Howard E. Ganther, Westport: Avi Publishing, Co, P. 176-191.

Welch, A.H., Lico, M.S., and Hughes, J.L. (1988). Arsenic in Ground Water in the Western Ground Water, 26(3), P. 333-347.

Witham, C. (2005). Volcanic disasters and incidents: a new database, Journal of Volcanology and Geothermal Research, 148, 191-233.

World Health Organisation. (2003). Guidelines for drinking water quality. Geneva

World Health Organisation. (2008). Guidelines for drinking water quality. Geneva

WRAP. (1999). The annual yearbook of the water resources management department. Water Resources Assessment Project, ed: Azza, N. G.T., Uganda. British Geological Survey 2001© NERC 2001.

APPENDIX

Appendix 1: joint readings

90
270
120
150
269
250
80
70
65
235
220
320
340
290
215
87
83
75
68
74
49
33
28
15
30

Appendix 2: Total hardness computation of the study area calculated from the equation $T = 2.497[(Ca^{2+}) + 4.11(mg^{2+})]$

Samples	Ca (mg/l)	Mg (mg/l)	Total Hardness (mg/l)	Remark
BH1	127.3	61.83	572	Very hard water
BH2	56.27	32.72	275	Very hard water
BH3	73.18	42.32	356	Very hard water
BH4	48.76	26.44	230	Very hard water
W1	22.27	13.45	111	Moderately hard water
W2	299.7	171.3	1452.2	Very hard water
W3	19.14	7.346	78	Moderately hard water
W4	31.9	14.84	141	hard water
W5	317.6	89.89	1163.5	Very hard water
W6	62.83	25.33	261	Very hard water
W7	74.85	37.58	341	Very hard water
W8	42.03	17.92	179	hard water
W9	46.05	28.47	232	Very hard water
W10	34.36	18.44	162	hard water
W11	52.54	36.52	281	Very hard water
W13	54.27	24.17	235	Very hard water
W14	254.8	99.17	1403	Very hard water
W16	26.4	9.325	104	Moderately hard water
W17	171	60.19	674	Very hard water
W18	282.5	139.9	1279	Very hard water
SW1	33.88	29.52	206	Very hard water
SW2	340.1	198.5	1665.9	Very hard water
SW3	53.06	27.12	244	Very hard water
SW5	67.83	31.8	300	Very hard water
SW6	85.57	52.08	427	Very hard water
SW7	67.77	35.54	315	Very hard water
SW8	75.39	46.54	379	Very hard water
SW9	17.57	7.749	75.7	Moderately hard water
SW10	16.92	6.879	70.5	Moderately hard water
SW11	32.56	16.56	149	hard water
			445.39333	Average total hardness

Appendix 3: Longitude and Latitude converted from minutes and seconds to degrees

Samples	Longitude	Latitude	Longitude	Latitude	Elevation (m)	Locality
BH 1	12°10'50.9"	10°38'00.3"	12.175	10.6	688	Waka
BH 2	12°09'48.1"	10°34'02.5"	12.158	10.567	750	Hema
BH 3	12°11'46.4"	10°35'49.5"	12.191	10.592	775	Ar. Barrack
BH 4	12°11'38.7"	10°36'17.4"	12.19	10.612	767	Biu
W 1	12°07'40.2"	10°36'48.4"	12.123	10.608	650	BCG
W 2	12°07'41.5"	10°36'57.3"	12.124	10.61	643	BCG
W 3	12°14'38.5"	10°31'55.8"	12.24	10.526	829	Yimirshika
W 4	12°13'07"	10°31'17.5"	12.228	10.52	837	Gwarta
W 5	12°12'40.3"	10°36'40.5"	12.206	10.608	711	Biu
W 6	12°12'47"	10°34'25.4"	12.207	10.571	770	Filin Jirgi
W 7	12°07'57.7"	10°35'35.4"	12.137	10.59	659	Biladega
W 8	12°08'22.8"	10°35'35.4"	12.139	10.589	681	Biladega
W 9	12°08'37.8"	10°33'31.4"	12.14	10.555	736	Tila
W 10	12°08'05.1"	10°35'51.6"	12.134	10.542	722	Tila
W 11	12°09'18.2"	10°33'13.3"	12.153	10.552	775	Tila
W 12	12°10'15.6"	10°37'11.5"	12.169	10.619	742	Tabra Fulani
W 13	12°09'47.3"	10°34'11.5"	12.158	10.569	723	Hema
W 14	12°09'46.6"	10°34'37.4"	12.158	10.573	727	Hema
W 15	12°12'04.2"	10°36'20"	12.201	10.603	772	Biu
W 16	12°12'29.8"	10°35'14.8"	12.205	10.586	793	Filin Jirgi
W 17	12°12'49.8"	10°35'	12.208	10.583	792	Filin Jirgi
W 18	12°12'27.6"	10°32'38.1"	12.205	10.54	759	Kunar
W 19	12°12'12.9"	10°31'20.1"	12.202	10.52	758	Ngwa
W 20	12°10'31.9"	10°33'36.6"	12.172	10.556	734	Tan
SW 1	12°08'00.9"	10°37'04.6"	12.134	10.618	644	BCG
SW 2	12°10'48.4"	10°37'49.7"	12.175	10.625	697	Waka
SW 3	12°11'44.1"	10°37'57.8"	12.191	10.626	670	Waka
SW 4	12°13'18.2"	10°37'23.1"	12.19	10.609	725	Biu
SW 5	12°08'03.2"	10°34'26.8"	12.22	10.621	722	Biu
SW 6	12°10'32.4"	10°36'20.9"	12.134	10.575	699	Hena
SW 7	12°09'55.6"	10°36'05.4"	12.172	10.604	725	Tabra Fulani
SW 8	12°10'48.1"	10°36'01.4"	12.157	10.601	700	Mallan
SW 9	12°10'52.9"	10°32'39.1"	12.175	10.6	747	Barrack
SW 10	12°10'41.8"	10°33'36.6"	12.176	10.54	758	Takwa
SW 11	12°11'39.2"	10°36'55.6"	12.174	10.555	729	Takwa